BEYOND TOMORROW

CUSTOMER OBSESSION IN STRATEGIC ALLIANCES

Learn Proven Strategies To Transform Your Strategic Alliances And Gain Competitive Advantage By Building Non-Linear Customer Obsessed Strategies.

This book is a must-read for anyone looking to navigate the complexities of strategic partnerships in today's dynamic market. It offers fresh perspectives on customer engagement and provides actionable insights and frameworks to enhance collaboration.

The author's extensive experience in fostering strategic alliances lends credibility to the concepts discussed, making it a valuable resource for both seasoned professionals and those new to the field.

By embracing the principles of customer obsession outlined in this book, readers can position themselves for sustained growth and innovation in their organizations.

"Strategic alliances are the backbone of global business today." — John C. Maxwel

Book Version 01-2024 November2024

Table of Contents

Dedication

I dedicate this book to my incredible Father-in-law, Saby D'Souza. His love, guidance, and practical wisdom have been an immense source of encouragement for me. His insatiable curiosity and commitment to growth, coupled with his attention to detail, have left a lasting impression on my life.

I can't help but think how proud he would be to see this book come to fruition. He exemplified the values of partnership, collaboration, faith, and respect for others—principles that shaped his life have shaped mine as well.

Thank you, Daddy, for your love and wisdom.

Preface

Welcome to *Customer Obsession In Strategic Alliances*! I'm thrilled to have you join me on this exciting journey through the ever-evolving landscape of strategic partnerships. As we navigate this dynamic environment, it's clear that industries are facing unprecedented disruptions. In light of this, the focus on customer obsession has never been more crucial.

In a world where traditional approaches often fall short, we must embrace a customer-obsessed mindset that empowers us to drive non-linear initiatives. This mindset encourages organizations to think creatively, respond with agility, and put customers at the very heart of their strategies.

Throughout this book, I'll share key insights on how to build strategic alliances that highlight the transformative power of customer obsession. My one hope is that as you dive deeper into this book, you will find inspiration to ignite fresh ideas to build critical capabilities to make strategic alliances thrive.

Why You Should Read This Book

1. **Gain a Fresh Perspective on Alliances**: Discover practical strategies to innovate and enhance customer engagement. Learn how to shift your organization's focus to a customer-obsessed model, ensuring that customer needs and feedback are at the forefront of your strategic initiatives.

2. **Gain actionable Insights:** To strengthen your partnerships and improve collaboration.

3. **Navigate Challenges Effectively**: Understand common pitfalls in strategic alliances and equip yourself with the tools needed to overcome these challenges, ensuring that your partnerships are resilient and effective.

4. **Drive Sustainable Growth**: Learn how to measure success through key performance indicators (KPIs) that align with both customer satisfaction and partnership goals, ultimately driving long-term growth for your organization.

I'm excited for you to join me on this transformative journey through customer-obsessed strategic alliances! Your adventure in discovering the boundless possibilities of strategic partnerships begins now!

Introduction

Welcome to an exciting journey into the world of strategic alliances! I'm thrilled to have you here as we embark on this exploration. In today's fast-paced and interconnected landscape, the ability to forge and sustain successful partnerships is more critical than ever. As organizations navigate the complexities of the modern business environment, those that master the art of collaboration will stand out from the competition, and I believe you're here to learn just how to do that.

Throughout this book, we'll dive into the key ingredients that contribute to thriving partnerships. Together, we'll discover how a shared vision, effective communication, and mutual benefits can create powerful synergies between organizations. It's not just about joining forces; it's about creating a harmonious relationship where both sides benefit and grow. Think of it as crafting a fine piece of music, each partner plays a vital role in creating a symphony that resonates with customers and drives success.

We'll explore essential concepts such as aligning goals, selecting the right partners, and nurturing a culture of innovation. Each chapter will offer you practical strategies, insights, and real-world examples that will guide you in applying these principles to your collaborations. But this isn't just a theoretical exercise; I encourage you to engage with the content actively. Reflect on your own partnerships: Are they truly aligned with your strategic objectives? Are you fostering open communication and trust? Are you ready to embrace change and adapt to new challenges?

Sneak peek at what each chapter holds:

Chapter 1: Understanding Strategic Alliances - In this chapter we will explore and learn at a high level the fascinating world of strategic alliances and why linear initiatives have failed in the past and focus our attention on to non-linear customer obsessed strategic alliance.

Chapter 2: The Pillars of Customer Obsessed Strategic Alliances - Explore how a customer-first mindset can transform your approach to partnerships. This chapter will challenge you to think about how well you really know your customers and how that knowledge can enhance your alliances.

Chapter 3: Deeper Dive Into Building Customer Obsessed Strategic Alliances -Effective communication can make or break a partnership. We'll dive into strategies for fostering open dialogue and building trust, and you'll learn how to avoid the common pitfalls that lead to misunderstandings.

Chapter 4 – The Equilateral Triangle: The Cornerstone of Strategic Alliances- In this chapter I present you an approach that has helped me build strong partnerships and focus on customer outcomes.

Chapter 5: Measuring Customer Success — The Right KPIs and Metrics - How do you know if your customer is realizing business value and is thriving? This chapter introduces customer-centric key performance indicators (KPIs) and associated metrics that will help you gauge success and make informed decisions.

Chapter 6: Measuring Strategic Alliance Success — The Right KPIs and Metrics

How do you know if your strategic alliance partnership is thriving? This chapter introduces you to strategic alliance specific key performance indicators (KPIs) and metrics that will help you gauge your progress and make informed decisions.

Chapter 7: Collaboration and Innovation in Strategic Alliances - This chapter focuses on the importance of developing a collaborative mindset. We will explore summaries of a few books that have shaped my personal point of view on accelerating innovation through strategic alliances.

Chapter 8: Overcoming Challenges in Partnerships - Every partnership faces obstacles. We'll explore common pitfalls and provide you with strategies to navigate these challenges effectively.

Chapter 9: Eight Pillars of Strategic Alliance - In this chapter I present to you a summary of eight pillars that are important to build a sustainable strategic alliances and lessons learned.

Chapter 10: Dynamic Coalition of Strategic Ecosystem – In this chapter I present to you a futuristic vision of strategic partnerships as a dynamic coalition of Ecosystem positioned for growth and market expansion.

Chapter 11: Self-Assessment and Reflection Guide on Strategic Alliances – In this chapter I present to you a self-assessment and reflection guide that will help you visualize your own journey towards maturing strategic alliances and build your action plan to making a difference.

Chapter 12: Conclusion and Key Takeaways

Imagine the impact of not just meeting customer expectations but exceeding them in ways that leave a lasting impression! Each page holds the potential for insights that could reshape your approach to partnerships and customer engagement.

Are you ready to take the first step into a world of non-linear customer obsessed strategic alliance, where collaboration leads to innovation and success? Let's get started!

Chapter 1

Understanding Strategic Alliances

Definition and Importance

Welcome to the fascinating world of strategic alliances! At its core, a strategic alliance is a formal agreement between two or more organizations to collaborate on a project or initiative while maintaining their independence. These partnerships leverage the unique strengths, resources, and capabilities of each partner to achieve common goals. In today's increasingly interconnected and rapidly changing business landscape, strategic alliances have emerged as essential tools for organizations seeking to innovate, expand their market reach, and enhance their competitive advantage.

The importance of strategic alliances cannot be overstated. In a world where challenges and opportunities abound, these collaborations allow organizations to pool resources, share risks, and access new markets or technologies more efficiently than they could achieve alone. For example, when Starbucks partnered with PepsiCo to market ready-to-drink coffee beverages, both companies were able to combine their strengths—Starbucks' brand equity and product innovation with PepsiCo's extensive distribution network. This partnership enabled Starbucks to expand its product line and market presence significantly.

Before we dive into the specific principles, it's important to understand that strategic alliances have evolved from simple cooperative agreements into complex, value-creating Ecosystem. Modern alliances require sophisticated management approaches that blend traditional partnership governance with advanced change management methodologies.

As we navigate the fast-paced and ever-changing landscape of modern business, we encounter a formidable challenge known as "the chasm." Coined by Geoffrey Moore in his influential work, "Crossing the Chasm," this concept highlights the gap between early adopters who eagerly embrace new ideas and the early majority who approach change more cautiously.

Understanding this chasm is essential for your organization if you aim to achieve widespread success and make a meaningful impact through strategic alliances.

Many organizations, perhaps including yours, embark on the exciting journey of forming strategic partnerships with high hopes and ambitious goals. The allure of collaboration often sweeps us into a whirlwind of enthusiasm, where the belief that a simple exchange of resources and knowledge will lead to groundbreaking innovations feels almost tangible. However, as many have experienced, this linear approach frequently leads to disappointment when the expected outcomes fail to materialize.

Imagine this scenario: two companies, inspired by the promise of innovation, join forces to develop a new product. They may possess complementary strengths and share a vision, yet if their strategy remains linear—solely focused on their internal objectives—they risk overlooking a critical element: the needs and desires of their customers. **This disconnect can create a gap between what is developed and what truly resonates in the market.**

When the product launches, it may find initial traction with early adopters, but without a robust strategy to engage the early majority, it struggles to gain momentum. The excitement that once surrounded the partnership begins to fade, leaving both organizations questioning where it all went wrong. It's a frustrating experience, one that feels disheartening, yet it is a common pitfall in today's business world. It is interesting to note that traditional, linear alliances have often fallen short in delivering the desired results.

The Need for Customer-Obsessed Strategic Engagements

Here's the good news: there is a powerful solution. By embracing a customer-obsessed approach to strategic engagements, your organization can bridge this chasm and create lasting connections with your audience. This mindset shifts the focus from internal goals to the very heart of your business—your customers.

Prioritizing customer obsession opens the door to deeper understanding and meaningful relationships. It requires you to actively listen to your customers, gather their insights, and involve them in the development process. This approach is not merely about asking for feedback; it's about creating a dialogue in which customers feel valued and heard. By genuinely understanding their needs, preferences, and pain points, you can co-create solutions that truly resonate with them.

Imagine the impact of collaborating with customers throughout the product development journey. Instead of working in isolation, you invite customers to share their experiences, suggest improvements, and help shape the final outcome. This collaborative spirit transforms your partnerships into dynamic ecosystems where innovation flourishes, benefiting everyone involved.

Consider the example of a tech company partnering with a healthcare provider to launch a digital health platform. By involving healthcare professionals and patients from the outset, the tech company gains invaluable insights that inform the design and functionality of the platform. The result is a solution that not only meets the expectations of early adopters but also resonates deeply with the early majority, laying a strong foundation for sustainable success.

The Ongoing Journey of Customer-obsessed Engagement

But the journey doesn't stop at the launch. Embracing a customer-obsessed approach means fostering ongoing relationships that extend beyond the initial rollout. By maintaining open lines of communication and continuously engaging with your customers, you empower your organization to adapt to their evolving needs and stay ahead of market trends.

As you reflect on your organization's approach to partnerships and product development, consider how you can cultivate a culture that prioritizes your customers. Embrace the opportunity to engage deeply with your audience, co-create meaningful solutions, and navigate the complexities of the market together. In doing so, you won't merely bridge the chasm; you'll forge a pathway for sustainable success and positive impact.

Remember, every interaction with your customers is an opportunity to build trust, foster loyalty, and drive innovation.

Here are the top ten reasons why alliances focused on linear initiatives have struggled to deliver meaningful results in the past:

1. **Rigid Structures:** Linear alliances often come with predetermined roles and responsibilities that limit flexibility. For instance, in the joint venture between DaimlerChrysler, the rigid structure and cultural differences led to significant challenges and ultimately, the dissolution of the partnership. (Source: "DaimlerChrysler: A Case Study in Mergers" by Michael D. Smith)

2. **Lack of Innovation**: These alliances can inhibit innovation as partners may stick to traditional processes rather than exploring new ideas. An example is the partnership between AOL and Time Warner, where the lack of innovation in integrating their platforms resulted in failure to capture the digital market. (Source: "The AOL-Time Warner Merger: A Cautionary Tale" by Fortune)

3. **Misalignment of Goals**: Linear partnerships may result in misaligned objectives between partners, leading to conflict and confusion. The alliance between General Motors and Toyota on the NUMMI plant struggled due to differing corporate cultures and goals. (Source: "NUMMI: A Case Study in Successful Joint Ventures" by Harvard Business Review)

4. **Ineffective Communication**: Poor communication can stifle collaboration and hinder progress. For instance, the partnership between Boeing and Lockheed Martin on the United Launch Alliance was marred by communication issues, affecting project timelines and deliverables. (Source: "Boeing and Lockheed Martin: A Partnership Complicated by Communication" by Aviation Week)

5. **Slow Response to Market Changes**: Linear alliances can be slow to adapt to market changes due to bureaucratic processes. The merger between Sprint and Nextel faced challenges as the companies struggled to integrate and respond to rapidly changing telecommunications markets. (Source: "Sprint and Nextel: A Merger Failure" by The Wall Street Journal)

6. **Overestimation of Synergies**: Partners may overestimate the synergies of their collaboration, leading to disappointment when those expectations are not met. The merger between HP and Compaq serves as a reminder of the pitfalls of overestimating potential synergies in a linear alliance. (Source: "HP and Compaq: Mergers and Acquisitions" by The Economist)

7. **Cultural Clashes**: Linear partnerships can amplify cultural differences that create friction and hinder collaboration. The merger between BBC and the independent production company left unresolved cultural tensions, ultimately limiting the effectiveness of their joint efforts. (Source: "BBC's Partnership Problems" by The Guardian)

8. **Inflexible Contracts**: Rigid contracts can lock partners into agreements that no longer serve their interests, preventing necessary adjustments. The partnership between Microsoft and Yahoo! was hampered by inflexible terms, stifling innovation and adaptability. (Source: "Microsoft and Yahoo: A Failing Partnership" by TechCrunch)

9. **Short-Term Focus**: Linear alliances may prioritize short-term gains over long-term success, leading to unsustainable practices. The partnership between Blockbuster and Enron to deliver video-on-demand services ultimately failed because it focused on immediate profits without considering long-term viability. (Source: "Blockbuster's Short-Sighted Decisions" by Forbes)

10. **Failure to Leverage Strengths**: Linear initiatives may overlook the unique strengths of each partner, leading to missed opportunities for growth. The

partnership between Kmart and Sears struggled to combine their strengths effectively, resulting in a decline in market presence. (Source: "The Kmart-Sears

Embracing a New Mindset

So, why should you challenge the status quo? The answer lies in the opportunity to adopt a new mindset that prioritizes customer obsession, commits to strategic alliances, and fosters strong collaboration. Imagine shifting from a linear, transactional approach to one that embraces flexibility and creativity—one that encourages you to think outside the box and respond proactively to customer insights.

As you reflect on your own organization, consider these questions:

- How can you shift from a linear approach to a more dynamic, customer-centric partnership?
- What strengths does your organization bring to the table that could be complemented by a potential ally?
- Are there industries outside your own that might provide unexpected insights or opportunities?

Types of Strategic Alliances

As we dive deeper into strategic alliances, it's important to recognize the various forms they can take. Understanding the types of strategic alliances will help you identify which model best suits your organization's objectives. Here are three key types:

1. Joint Ventures

Joint ventures involve the creation of a separate legal entity where both partners contribute resources and share in profits and risks. This collaborative approach can yield remarkable results when partners combine their unique strengths. A prime example is the partnership between Sony and Ericsson, which birthed Sony Ericsson. By merging their expertise in technology and telecommunications, they developed innovative mobile solutions that captured significant market share.

In a joint venture, partners are not just allies but co-creators, allowing them to tackle larger projects and enter new markets with confidence. This model can lead to groundbreaking products and services that benefit from the combined capabilities of both organizations.

2. Equity Partnerships

In an equity partnership, one company takes an ownership stake in another. This type of alliance fosters a deeper commitment to collaboration and encourages partners to align their strategic interests more closely. A well-known example of this is the partnership between Coca-Cola and Monster Beverage Corporation. By acquiring a significant equity stake in Monster, Coca-Cola was able to tap into the burgeoning energy drink market, diversifying its portfolio and reaching new consumer segments.

Equity partnerships can provide stability and a shared vision, making them a powerful avenue for long-term growth and innovation. They allow partners to work closely together, ensuring that their goals and values align as they navigate the complexities of the market.

3. Non-Equity Collaborations

Non-equity collaborations are often less formal than the previous two types and involve contractual agreements between organizations to work together on specific projects or initiatives without exchanging equity. These alliances can be incredibly flexible, allowing partners to focus on particular objectives without the complexities of ownership stakes.

A notable example is the partnership between Microsoft and Adobe. By integrating their software products, they created a more seamless user experience for customers. This non-equity collaboration has allowed both companies to benefit from each other's technologies, enhancing customer satisfaction and loyalty.

Embracing a Customer-Obsessed Mindset

As we explore these various forms of strategic alliances, it's crucial to emphasize the importance of a customer-obsessed mindset. In *Beyond Tomorrow*, we advocate for organizations to place customers at the center of their strategies. By doing so, partners can address emerging customer needs more effectively and creatively.

For example, in a joint venture between two tech companies, understanding customer feedback and preferences can guide product development and marketing strategies, ensuring that the resulting solutions resonate with the target audience. Similarly, in equity partnerships, collaborative efforts should focus on enhancing customer experiences, driving loyalty, and creating lasting relationships.

As we continue through *Beyond Tomorrow*, we will explore how adopting agile frameworks, creating shared innovation labs, building dynamic Ecosystem, and leveraging collective agility can empower your organization to thrive amidst disruption. By embracing these non-linear approaches, we can reimagine partnerships that prioritize customer satisfaction, drive innovation, and create lasting value.

Summary

In this chapter, we have laid the foundation for understanding strategic alliances and their importance in today's business landscape. Whether through joint ventures, equity partnerships, or non-equity collaborations, the potential for growth and innovation is immense.

Now, as you move forward in this book, keep these insights in mind. There is much more to explore about how to develop effective strategic alliances that thrive on customer obsession.

Turn the page to the next chapter, where we will dive deeper into the foundations of strategic alliances, equipping you with the knowledge and tools needed to succeed. Your journey towards creating meaningful partnerships that resonate with your audience begins here—let's continue this exploration together!

What changes can you make to embrace a more customer-centric approach? How can you encourage your teams to think innovatively about their collaborations?

The future of strategic alliances awaits!

"Alone we can do so little; together we can do so much." — Helen Keller

Chapter 2

The Pillars of Customer Obsession in Strategic Alliances

I'm excited to dive deep into how a customer-first mindset can transform your approach to partnerships in today's dynamic business landscape. As organizations seek to enhance their competitive edge, the ability to forge and sustain successful strategic alliances has never been more critical. However, this endeavor requires much more than just a handshake; it demands careful planning, open communication, and a genuine commitment to mutual goals.

At the heart of this discussion is the recognition that relationships form the very foundation of business success. As highlighted in the previous chapter, strategic alliances can thrive when partners embrace a customer-obsessed mindset. Technology has made it easier than ever to connect, creating a "digital concord" where buyers and sellers can engage meaningfully. As we explore the pillars of customer obsession, we'll uncover not only the critical performance indicators for successful partnerships but also the secrets to thriving alliances. We'll tackle common challenges that can lead to failures and identify what I believe to be the most telling key performance indicator (KPI) of success in these collaborations.

Clear Alignment of Goals

At the core of every successful alliance is a clear understanding of shared objectives. It's essential to sit down with your potential partners and articulate what you hope to achieve together. This process involves not just outlining high-level goals but diving into the specifics, such as market entry strategies or product development plans.

When it comes to customer obsession, aligning your goals with customer needs is paramount. For instance, if your partnership aims to launch a new product, consider these questions: How does this product solve a customer problem? What feedback have we gathered from customers that informs us of this goal? How well do your partnership goals align with your organization's strategic objectives?

Let's take a look at two companies in the automotive sector that joined forces to develop electric vehicles. By aligning their goals around sustainability and customer preferences for eco-friendly options, they not only achieved their internal targets but also positioned themselves favorably in the eyes of environmentally conscious consumers. This illustrates how focusing on customer needs can drive shared objectives and enhance collaboration.

As you reflect on your partnerships, consider the specific goals you share with your partners and how these objectives align with the needs of your customers. It's essential to take a moment to think about whether your shared goals truly resonate with what your customers are looking for. Are you consistently revisiting and adjusting these goals based on the feedback you receive from your customers? This practice is crucial, as it ensures that you remain responsive to their changing preferences and expectations.

Additionally, think about how effectively your objectives are communicated to all stakeholders involved in the partnership. Clear communication fosters alignment and understanding, which are vital for the success of any collaboration. Ask yourself: in what ways do your shared goals create tangible value for your customers? Understanding this connection will help strengthen your partnerships and enhance the overall customer experience.

Lastly, consider how you might better integrate customer insights into your goal-setting process. This integration can lead to more informed decisions and innovative approaches, ultimately driving your partnership's success and ensuring that it remains centered around the needs of your customers. Taking the time to reflect on these aspects will empower you to cultivate stronger alliances that thrive in today's competitive landscape.

Thorough Partner Selection

Choosing the right business partner can make or break your alliance. This critical step requires time and thorough research to gather the necessary data, helping create a "platform of mutual capabilities and scale." Don't rush this process; it's worth taking the time to assess potential partners based on their strengths, resources, and cultural fit.

When applying a customer-obsessed mindset, it's crucial to evaluate potential partners through the lens of customer needs. A partner that excels in customer service can significantly enhance your ability to deliver exceptional experiences. Look for organizations that complement your own, filling gaps and enhancing your competitive position.

For example, a technology firm that partnered with a customer service solutions provider improved its service delivery by selecting a partner known for its excellent customer support. This decision resulted in elevated customer satisfaction and retention rates, showcasing the power of thorough partner selection.

As you contemplate the potential partners for your alliances, it's important to consider the criteria you use for evaluation. Are these criteria comprehensive enough to include a focus on customer service? Understanding how a potential partner prioritizes customer satisfaction can significantly influence the success of your collaboration.

Next, think about how you assess a partner's alignment with your organization's customer-centric values. It's essential that both parties share a commitment to putting customers first, as this shared ethos will be crucial in driving joint initiatives.

Equally important is the diligence with which you conduct your research on potential partners. Are you taking the time to perform thorough due diligence, and what additional investigations could enhance your selection process? The more informed you are about a potential partner's operations, culture, and market position, the better equipped you'll be to make a strategic decision.

Additionally, reflect on how well you understand your partner's capabilities and strengths in serving customers. Having a clear picture of what your partner can offer will help you identify how their strengths can complement your own and fill any gaps in your service delivery.

Finally, consider the mechanisms you have in place to monitor the effectiveness of your partner relationships. Establishing clear metrics and regular check-ins can help ensure that both parties remain aligned and responsive to customer needs as the partnership evolves. By thoughtfully addressing these areas, you can set the stage for a successful, customer-obsessed strategic alliance.

Open Communication

It's often said that good communication builds strong relationships. Establishing a culture of open and honest dialogue is essential for any strategic alliance that aims to withstand the test of time. Regular check-ins, updates, and feedback loops are critical for building trust and ensuring both parties remain aligned.

In the context of customer obsession, fostering open communication extends beyond your partnership; it should also include your customers. Regularly soliciting customer feedback will help ensure that the partnership continues to meet customer needs. Consider how you can foster a more open dialogue in your current collaborations.

For example, a marketing agency that worked closely with a client held weekly meetings to discuss ongoing campaigns. This transparency allowed them to pivot quickly based on customer feedback, leading to campaigns that resonated more effectively with their target audience. Such an approach not only strengthens the partnership but also enhances customer engagement.

As you consider the dynamics of your partnerships, it's crucial to evaluate how frequently you communicate with your partners. Is the frequency sufficient to keep everyone aligned and informed? Open lines of communication are vital for ensuring that all parties are on the same page and working toward common goals.

Additionally, reflect on whether there are established channels for feedback between your organization and your partners. Creating a structured way for both sides to exchange insights and concerns can significantly enhance collaboration.

Think about how you currently incorporate customer feedback into your communications with partners. Are you actively sharing relevant customer insights that can inform you of your joint strategies? This practice not only strengthens your partnerships but also ensures that decisions are rooted in a deep understanding of customer needs.

It's also important to identify any barriers that may hinder open communication. Are there specific challenges that create obstacles in your interactions? Acknowledging these barriers is the first step toward overcoming them, allowing for a more transparent and effective partnership.

Finally, consider whether you are proactive in sharing both successes and challenges with your partners. Fostering an environment of transparency can build trust and strengthen your alliances, making it easier to navigate obstacles together. By addressing these areas, you can cultivate a partnership that thrives on open communication and mutual support.

Structured Governance

You've probably heard the saying, "What you cannot measure, you cannot manage." Implementing a robust governance structure provides clarity in roles and responsibilities, decision-making authority, and conflict resolution processes. This structure serves as the backbone of the alliance you seek to build, grow, and mature.

A well-defined governance framework is essential for aligning with customer obsession. Establishing clear guidelines helps partners prioritize customer needs in decision-making. Does your current partnership have a structured governance framework? If not, what steps can you take to implement one?

For instance, a healthcare provider that partnered with a technology firm established a governance structure that included regular performance reviews based on customer satisfaction metrics. By keeping customers at the center of their discussions, they ensured that their services continuously improved and adapted to customer expectations.

As you evaluate the effectiveness of your partnerships, it's essential to consider whether you have a structured governance framework in place. Does this framework include customer-focused metrics that guide decision-making and performance evaluations? A robust governance structure is crucial for ensuring that both parties remain aligned with their customer-centric goals.

Next, reflect on how clearly defined the roles and responsibilities are within your partnership. Clear delineation of responsibilities helps prevent misunderstandings and ensures that everyone knows their contributions toward achieving shared objectives.

Additionally, think about how regularly you review your performance in relation to customer satisfaction. Are you taking the time to assess how well your joint efforts are meeting customer needs? This regular review process is vital for identifying areas for improvement and ensuring that both partners are committed to enhancing the customer experience.

When conflicts arise during collaboration, consider how effectively you resolve them. Open and constructive conflict resolution can strengthen relationships and prevent issues from escalating. Reflect on the mechanisms you have in place for conflict resolution and how they contribute to maintaining a positive partnership dynamic.

Finally, contemplate what improvements can be made to your governance structure to better serve your customers. Are there adjustments or enhancements that could lead to more effective collaboration and ultimately create greater value for your customers? By thoughtfully addressing these aspects, you can strengthen your governance framework and ensure that your partnerships are positioned for success in meeting customer expectations.

Focus on Relationship Management

Think big and consider your alliance as a long-term relationship rather than a one-off transaction. Investing time and resources in building and nurturing that connection is vital. Engage in team-building activities, joint workshops, and collaborative problem-solving sessions to strengthen your partnership.

This commitment to relationship management ties directly back to customer obsession. When partners are deeply connected, they can better anticipate and respond to customer needs, creating an environment where collaboration flourishes. How can you deepen the connections within your partnerships?

For example, an international retailer that regularly hosted joint workshops with their supply chain partners strengthened their relationship and led to innovative inventory solutions that ultimately enhanced customer satisfaction. Such collaborative efforts can be the key to unlocking additional value for your customers.

As you think about the strength of your partnerships, consider how much time you dedicate to relationship-building activities with your partners. Investing in these connections is vital for fostering trust and collaboration, so it's worth evaluating whether you're allocating sufficient resources to this important aspect of your alliance.

Reflect on the collaborative initiatives you have undertaken to strengthen your partnership. What specific actions have you taken to enhance cooperation and mutual support? These initiatives can play a significant role in reinforcing your commitment to one another and in achieving shared goals.

Additionally, consider whether you are actively fostering a culture of teamwork and shared success within your alliances. A partnership thrives when both parties feel a sense of unity and purpose. Are your team members encouraged to work together, share insights, and celebrate collective achievements?

It's also essential to ensure that all stakeholders are engaged and invested in the partnership. How do you keep everyone informed and motivated? Engaging stakeholders at all levels can enhance alignment and commitment, leading to more effective collaboration.

Finally, think about the steps you can take to create more opportunities for meaningful interactions with your partners. Are there specific events, workshops, or brainstorming sessions that could facilitate deeper connections? By prioritizing these interactions, you can build a stronger foundation for collaboration and drive greater value for both organizations.

Flexibility and Adaptability

In today's rapidly changing business environment, the ability to adapt is more crucial than ever. Markets shift, technologies advance, and customer needs evolve. Therefore, organizations must remain open to reassessing their strategies and aligning their goals with these changes.

Being flexible and adaptable means rethinking your approach when circumstances change. It involves regularly evaluating how well your partnership is functioning and making necessary adjustments to stay aligned with each partner's objectives. This adaptability is vital to ensure that customer needs remain at the forefront. Are you prepared to embrace change when it comes your way?

During the pandemic, many companies had to pivot their strategies quickly. For instance, a restaurant chain partnered with a delivery service to meet changing customer preferences. This flexibility not only helped retain their existing customer base but also expanded it by reaching new customers who preferred delivery options.

As you reflect on the agility of your partnerships, consider how quickly your organization can pivot its strategies in response to changing market conditions. In today's fast-paced environment, the ability to adapt swiftly is crucial for maintaining a competitive edge. Are you prepared to make necessary adjustments when circumstances demand?

It's also important to evaluate whether you are regularly reviewing your partnership's performance to identify areas for adjustment. Conducting these reviews allows you to assess how well your collaboration is meeting its objectives and whether any shifts are needed to enhance effectiveness.

Think about how well you communicate changes and expectations with your partners. Open and transparent communication is vital for ensuring that all parties are aligned and informed. Are you proactively sharing updates and discussing adjustments in strategy? This practice can help prevent misunderstandings and foster a more cohesive working relationship.

Consider the processes you have in place to facilitate agile decision-making. Are there mechanisms that enable your team to respond promptly to challenges or opportunities? By streamlining decision-making, you can enhance your partnership's responsiveness and ability to navigate uncertainties.

Finally, reflect on how you can encourage a culture of adaptability within your partnership. Are you fostering an environment where both organizations feel empowered to experiment and adjust as needed? By promoting flexibility and openness to change, you can build a partnership that thrives in the face of evolving market dynamics. Embracing these elements will help ensure that your alliances remain resilient and effective, ultimately benefiting both organizations and the customers you serve.

Culture of Innovation

Encouraging a culture of innovation is essential for fostering adaptability. When both partners feel empowered to think creatively and propose new ideas, they are more likely to find effective solutions to emerging challenges. This culture should promote open communication and support for experimentation, allowing teams to explore different approaches without fear of failure.

A focus on innovation aligns closely with customer obsession. By fostering a mindset open to experimentation, organizations can develop new solutions that respond directly to customer feedback and changing needs. Are you ready to embrace change? How can you cultivate an environment that encourages flexibility and innovation?

To foster a robust culture of innovation, consider implementing the following strategies:

- **Encourage Diverse Perspectives:** Invite team members from various backgrounds and departments to contribute ideas. This diversity fosters creativity and leads to unique solutions.
- **Establish Innovation Labs**: Create dedicated spaces for brainstorming and experimentation, where employees can freely explore new concepts without the pressure of immediate results.
- **Reward Innovation**: Recognize and celebrate innovative ideas and successful implementations, reinforcing the value of creativity within the organization.
- **Emphasize Learning**: Promote a mindset that views failures as learning opportunities, encouraging teams to analyze what didn't work and how to improve moving forward.

For example, a tech company implemented a program allowing employees to dedicate a portion of their time to exploring innovative ideas. This initiative led to the development of a new feature that dramatically improved user engagement and customer satisfaction, showcasing how a culture of innovation can directly impact customer experience.

As you consider the role of innovation in your partnerships, reflect on how you currently encourage innovative thinking within your organization and with your partners. Are you actively fostering an environment that prioritizes creativity and exploration? Encouraging innovation is essential for staying competitive and responding effectively to customer needs.

Next, think about whether you are creating an environment where team members feel comfortable sharing their ideas. A culture that supports open dialogue and values contributions from all individuals can lead to groundbreaking solutions. Are your team members empowered to voice their thoughts and suggestions without fear of judgment?

Consider what specific steps you can take to provide resources for experimentation and creativity. This might include allocating time for brainstorming sessions, investing in training programs, or establishing innovation labs where employees can explore new concepts. Providing the right resources not only stimulates creativity but also demonstrates a commitment to innovation.

Additionally, evaluate how you celebrate and learn from both successes and failures in your efforts to accelerate innovation. Recognizing achievements reinforces positive behaviors, while analyzing setbacks can provide valuable lessons for future initiatives. Are you taking the time to acknowledge milestones and discuss what didn't work, fostering a culture of continuous improvement?

Finally, contemplate what new ideas or features you could explore that would directly address your customers' needs. Engaging with customers to understand their preferences and pain points can inspire innovative solutions that enhance their experience. By prioritizing these elements, you can cultivate a vibrant culture of innovation that drives meaningful value for your partnerships and the customers you serve.

Summary

As we explore the pillars of customer obsession in strategic alliances, it becomes clear that building trust, aligning goals, maintaining open communication, establishing structured governance, managing relationships effectively, being flexible, and fostering innovation are all critical elements. Each of these pillars contributes to creating alliances that not only survive but thrive in the face of challenges.

In the previous chapter, we discussed how linear initiatives often fail to deliver results due to rigid structures, misaligned goals, and a lack of innovation. As you reflect on your own partnerships, consider where you might improve. By committing to these principles and maintaining a customer-centric focus, you can create strategic alliances that empower both organizations and deliver meaningful value to your customers.

As we move forward in this book, let's carry these insights into our next chapter, where we will further explore customer-obsessed strategic alliances. Together, we can unlock the potential for innovation and collaboration, driving meaningful results that resonate across industries.

Are you ready to take the next step? I encourage you to take the self-reflection questions and associated scoring mechanisms to build a personal action plan.

Self-Reflection Questions on the Pillars of Customer Obsession in Strategic Alliances

These fifteen questions are designed to help you assess your current practices, align your strategies, and identify areas for improvement. For each question, you can use a simple scoring mechanism to rate your responses on a scale of 1 to 5, with 1 being "strongly disagree" and 5 being "strongly agree." This will help you gauge your current standing and plan actionable next steps.

Self-Reflection Questions

1. **Goal Alignment**: We have clearly defined shared objectives with our partners that align with customer needs.
2. **Partner Evaluation**: Our criteria for evaluating potential partners include a focus on customer service.
3. **Diligence**: We conduct thorough due diligence to understand our partners' strengths, resources, and cultural fit.
4. **Communication**: We communicate our shared goals and objectives effectively to all stakeholders involved in the partnership.
5. **Feedback Integration**: We regularly integrate customer feedback into our goal-setting process.
6. **Performance Reviews**: We consistently review our partnership's performance in relation to customer satisfaction metrics.
7. **Conflict Resolution**: We have effective processes in place for resolving conflicts that arise during our collaboration.
8. **Relationship Building**: We dedicate adequate time to relationship-building activities with our partners.
9. **Culture of Teamwork**: Our organization fosters a culture of teamwork and shared success within our alliances.
10. **Flexibility**: We can pivot our strategies quickly in response to changing market conditions.
11. **Innovation Encouragement**: We actively encourage innovation within our organization and with our partners.

12. **Open Communication**: We maintain open lines of communication with our partners to facilitate trust and transparency.
13. **Engagement**: All stakeholders are engaged and invested in the partnership.
14. **Monitoring Effectiveness**: We have mechanisms in place to monitor the effectiveness of our partner relationships.
15. **Adaptability**: We cultivate a culture of adaptability within our partnership to respond to new challenges.

Use The Scoring Mechanism

For each question, assign a score from 1 to 5 based on your agreement with the statement, and calculate your total score and categorize it as follows:

- **15-30**: Significant Improvement Needed
- **31-45**: Some Areas for Development
- **46-60**: Good Practices in Place
- **61-75**: Strong Alignment and Practices
- **76-90**: Excellent Practices and Leadership in Customer Obsession

Build An Action Plan By Using The Score Guide

- **For scores 15-30**: Focus on foundational elements. Identify specific areas where you need to develop structures or practices. Consider engaging in workshops to align your team and partners around customer-centric goals.

- **For scores 31-45**: Identify key priorities for improvement. Create a roadmap to enhance communication and collaboration with partners. Schedule regular performance reviews and integrate more customer feedback into your processes.

- **For scores 46-60**: Build on your strengths while addressing areas for improvement. Invest in relationship-building activities and training for your teams to enhance customer service capabilities.

- **For scores 61-75**: Continue to refine your practices and share successful strategies with your partners. Lead initiatives to foster innovation and adaptability across your partnerships.

- **For scores 76-90**: Celebrate your successes and take on a mentorship role with other organizations seeking to enhance their customer-obsessed practices. Consider sharing your insights through case studies or presentations at industry events.

	Use this table to record your scores	Score Between 1-5
1	**Goal Alignment**: We have clearly defined shared objectives with our partners that align with customer needs.	
2	**Partner Evaluation**: Our criteria for evaluating potential partners include a focus on customer service.	
3	**Diligence**: We conduct thorough due diligence to understand our partners' strengths, resources, and cultural fit.	
4	**Communication**: We communicate our shared goals and objectives effectively to all stakeholders involved in the partnership.	
5	**Feedback Integration**: We regularly integrate customer feedback into our goal-setting process.	
6	**Performance Reviews**: We consistently review our partnership's performance in relation to customer satisfaction metrics.	
7	**Conflict Resolution**: We have effective processes in place for resolving conflicts that arise during our collaboration.	
8	**Relationship Building**: We dedicate adequate time to relationship-building activities with our partners.	
9	**Culture of Teamwork**: Our organization fosters a culture of teamwork and shared success within our alliances.	
10	**Flexibility**: We can pivot our strategies quickly in response to changing market conditions.	
11	**Innovation Encouragement**: We actively encourage innovation within our organization and with our partners.	
12	**Open Communication**: We maintain open lines of communication with our partners to facilitate trust and transparency.	
13	**Engagement**: All stakeholders are engaged and invested in the partnership.	
14	**Monitoring Effectiveness**: We have mechanisms in place to monitor the effectiveness of our partner relationships.	
15	**Adaptability**: We cultivate a culture of adaptability within our partnership to respond to new challenges.	
	Add your total score and build your action plan	

Chapter 3:

A Deeper Dive Into Customer-Obsessed Strategic Alliance

Welcome to an exciting and transformative exploration of customer obsession an approach that transcends standard business strategies and reshapes how organizations operate. In this chapter, we will delve deeply into the essential strategies and practices that foster a customer-obsessed mindset within strategic alliances. By aligning partnership goals with customer needs, we can pave the way for meaningful collaboration and innovative solutions that resonate profoundly with clients.

The Essence of Customer Obsession

In a fast-paced and ever-evolving marketplace, customer obsession is not merely a buzzword or a fleeting trend to add to your company's repertoire; it is a fundamental philosophy that can redefine organizational interactions with their clients. The shift from merely understanding customer needs to effectively implementing strategies that address those needs is crucial for success. This alignment of understanding and execution creates a robust framework, particularly within strategic alliances where partners must work collaboratively to deliver exceptional value.

Imagine a world where every decision, whether major or minor, is driven by the genuine needs and desires of customers. Picture organizations that do not simply aim to meet expectations but actively pursue opportunities to exceed them in delightful and unexpected ways. This proactive approach fosters customer loyalty and transforms ordinary transactions into extraordinary experiences. By wholeheartedly embracing customer obsession, organizations can cultivate deeper relationships with their clients, ultimately creating a competitive edge that is difficult to replicate.

Understanding Customer Needs

At the core of customer obsession lies a profound understanding of what customers truly want and need. This goes far beyond simply gathering data or conducting surveys; it requires a sincere commitment to listening to the voice of the customer and integrating that feedback into every aspect of the business.

Consider your customers as partners in this journey—when you truly listen to them, you are empowered to create products and services that resonate authentically and meaningfully.

A critical first step in this journey is actively seeking customer feedback through diverse channels, be it surveys, focus groups, or social media interactions. Think of customer feedback as a treasure map; it can lead you directly to valuable insights that help refine your offerings and elevate the customer experience.

For example, consider a tech company preparing to launch an innovative new app. Instead of merely rolling it out and waiting for reactions, the company decides to engage potential users before the official launch. By hosting focus groups and beta testing sessions, they gather invaluable insights that highlight specific user frustrations and desires. The resulting adjustments not only enhance the app's user experience but also create a product that customers rave about upon launch. This proactive approach exemplifies how understanding and responding to customer needs can lead to strong brand loyalty and satisfaction.

Creating Value for Customers

While understanding customer needs is the first step, the real magic happens when organizations effectively execute those insights. This is where the concept of premium execution comes into play. Renowned thought leader Ram Charan emphasizes that even the best ideas can falter without effective implementation. To genuinely embrace customer obsession, organizations must align every aspect of their operations—from leadership to frontline employees—around the common goal of delivering exceptional value to customers.

Creating value extends beyond simply providing a product or service; it encompasses curating an entire experience that meets and anticipates customer needs. This requires an ongoing dialogue with customers, allowing companies to adapt and innovate continuously. A customer-centric approach should permeate every level of the organization, ensuring that every employee understands their role in enhancing the customer journey.

Consider a global retail chain that embarks on a significant transformation to become more customer centric. They recognize the importance of integrating feedback at every operational layer and decide to invest in comprehensive training programs designed to instill a customer-first mindset among employees. As a result, customer satisfaction scores soar, and employee morale improves dramatically. Staff members feel more connected to their work and understand the value they bring to customers.

This alignment is akin to a well-oiled machine, where every cog works harmoniously to ensure a seamless experience for the end user. The synergy between individual contributions and overarching customer goals not only enhances the customer journey but also fosters a culture of empowerment and engagement among employees.

The Boundaryless Customer Engagement In Strategic Alliances

In today's interconnected world, the concept of **boundaryless customer engagement** emerges as a transformative force, playing a crucial role in understanding and addressing customer needs. At its core, boundaryless engagement refers to the ability of organizations to operate without traditional barriers, allowing for fluid and dynamic interactions across multiple channels and touchpoints. This approach is essential for embracing non-linear initiatives that foster collaboration and innovation, enabling companies to respond more effectively to the complex and evolving expectations of their customers.

Imagine the exciting possibilities that arise when strategic alliances empower organizations to extend their reach and capabilities. By forming these dynamic partnerships, companies can share invaluable insights and resources, paving the way for a deeper and more comprehensive understanding of customer behaviors and preferences. Instead of being confined to rigid frameworks, organizations can navigate the intricate pathways of customer engagement, creating an ecosystem where insights flow freely, and collaboration thrives.

Picture a technology company joining forces with a leading data analytics firm. This partnership represents the essence of boundaryless engagement, where both organizations pool their expertise to create a unified platform that captures customer interactions across various channels, be it online, in-store, or through customer service. This collaborative effort not only enhances the customer journey but also enables both organizations to gain richer insights into customer behaviors, allowing them to refine their offerings in real time.

Such boundaryless engagement exemplifies non-linear strategic initiatives, where organizations bravely venture beyond traditional silos and embrace an interconnected landscape. In this vibrant environment, every partner contributes its unique strengths, resulting in a synergistic effect that elevates the entire customer experience. Businesses can now respond nimbly to customer feedback, ensuring that their strategies remain not only relevant but also deeply impactful.

Consider a cloud service provider that collaborates with an AI firm to revolutionize customer engagement tools. By harnessing advanced AI algorithms, these companies analyze vast amounts of customer interactions, identifying patterns and predicting needs with remarkable accuracy. This boundaryless engagement creates a seamless flow of information, allowing organizations to respond swiftly to customer feedback and deliver personalized experiences that truly resonate.

The beauty of boundaryless engagement lies in its ability to unlock new potential. Imagine a scenario where organizations not only listen to their customers but also anticipate their needs and desires. This proactive approach redefines the customer experience, transforming it from a transactional interaction into a genuine relationship, where customers feel valued and understood.

As we continue to delve deeper into the concept of boundaryless customer engagement through strategic alliances, it becomes evident that this innovative approach is essential for navigating the complexities of today's market. It invites organizations to rethink how they collaborate and engage with their customers, crafting an exhilarating environment ripe for growth and innovation.

Together, with your service providers you can harness the power of strategic partnerships to create a future where customer obsession reigns supreme, and every interaction leaves a lasting impression.

Embrace this vision! The possibilities are limitless when organizations unite to create value for their customers, transcending boundaries and redefining what it means to engage meaningfully. I invite you to embark on this thrilling journey to unlock the extraordinary potential of customer obsession through boundaryless engagement!

Proactive Listening and Adaptation

To genuinely grasp customer needs, organizations must actively seek feedback through various channels. This includes crafting well-structured surveys to gather quantitative data, hosting focus groups for qualitative insights, and engaging in real-time conversations on social media platforms. Each of these channels offers unique opportunities for understanding the voice of the customer.

For example, after facing declining user engagement, a software company analyzed user feedback and discovered that many found the onboarding process too complex. This revelation served as a turning point; by simplifying the onboarding experience, they could significantly boost user retention and satisfaction. In this case, customer feedback acted as a treasure map, guiding the organization toward valuable improvements.

Fostering a Culture of Innovation

Encouraging a culture of innovation is essential for fostering adaptability within strategic alliances. When both partners feel empowered to think creatively and propose new ideas, they are more likely to identify effective solutions to emerging challenges. This culture should promote open communication and support for experimentation, allowing teams to explore different approaches without fear of failure.

To cultivate a robust culture of innovation, organizations can take several strategic steps. First, they should encourage diverse perspectives by inviting team members from various backgrounds and departments to contribute ideas. This diversity fosters creativity and leads to unique solutions. Establishing innovation labs can create dedicated spaces for brainstorming and experimentation, where employees can explore new concepts without the pressure of immediate results. Recognizing and celebrating innovative ideas reinforces the value of creativity within the organization and promotes a mindset that views failures as learning opportunities.

For instance, a tech company implemented a program allowing employees to dedicate a portion of their time to exploring innovative ideas. This initiative led to the development of new features that dramatically improved user engagement and customer satisfaction, showcasing how a culture of innovation can directly impact the customer experience.

Reimagining Innovation Beyond The Straight Line

Imagine for a moment that innovation is not just a straightforward journey from point A to point B but a vibrant, interconnected ecosystem brimming with possibilities. This is the essence of non-linear initiatives—a revolutionary approach that transforms how you, as an organization, navigate complexity, uncertainty, and the rapid evolution of technology. As you embark on this exploration, prepare to unlock new perspectives and delve into the exciting world of customer-obsessed strategic partnerships.

A Journey into Transformation: TechNova Semiconductor

Let's begin by considering your potential path to transformation, drawing inspiration from the story of TechNova, a mid-sized semiconductor manufacturer that found itself ensnared in the confines of linear thinking. Their product development cycle mirrored a rigid assembly line, with each stage—concept, design, prototype, testing, and production—meticulously planned but frustratingly slow. Innovation felt like a distant dream, lost amidst a sea of disconnected processes, leaving customer needs unmet and opportunities overlooked.

However, everything changed in 2022 when CEO Elena Rodriguez ignited a spark of change by introducing the "Adaptive Nexus Strategy." This was not merely a new methodology; it was an invitation for you to think differently about how to serve customers better.

Picture TechNova's transformation as they dismantled traditional departmental boundaries, creating dynamic cross-functional "innovation cells." Imagine researchers, engineers, marketers, and customer experience specialists collaborating in real time on a digital platform that facilitated instantaneous knowledge sharing. This collaborative ecosystem breathed new life into the company, enabling it to swiftly respond to market shifts and evolving customer needs.

But the metamorphosis didn't stop there. TechNova embraced a **Distributed Intelligence Framework**, harnessing the power of AI to continuously analyze market signals. Teams were empowered to pivot instantly based on emerging technological trends, embracing a "fail-fast, learn-faster" culture that celebrated experimentation. The results? A stunning reduction in product development time by an impressive 62% and an increase in the innovation hit rate from 18% to an astonishing 47%. TechNova created three breakthrough semiconductor designs that disrupted existing market architectures, demonstrating that with the right mindset, innovation knows no bounds and can be relentlessly focused on customer satisfaction.

A New Approach: GlobalSync Systems

Now, let's shift our focus to another exemplary story: GlobalSync, a global systems integrator that traditionally operated through a linear, sequential service delivery model. Clients would specify their requirements, and teams would methodically implement solutions, often leading to misalignment and delayed value realization. But under the visionary leadership of CTO Marcus Chen, GlobalSync embraced a radical transformation known as the "Adaptive Systems Orchestration" approach, redefining their commitment to customer engagement.

Envision a world where static project blueprints are replaced with adaptive, real-time solution frameworks. Imagine machine learning algorithms recalibrating project parameters on the fly, creating "solution Ecosystem" instead of fixed deliverables. This dynamic approach allowed GlobalSync to redefine service integration in ways that thrilled clients and exceeded their expectations.

By employing **Continuous Value Mapping**, GlobalSync developed advanced telemetry systems that tracked client value in real time. This capability enabled instant course corrections based on emerging client needs, transforming service contracts from fixed deliverables into dynamic value generation agreements. The results were extraordinary: client satisfaction scores surged by 78%, project implementation times decreased by 55%, and the company generated three times more innovation opportunities per client engagement, all rooted in a deep commitment to understanding and serving customer needs.

The Philosophical Shift: Why Non-Linear Matters

But why should you care about non-linear initiatives in the context of customer-obsessed strategic alliances? They represent more than just a tactical shift—they embody a profound philosophical transformation. They encourage you to move from control to adaptability, from prediction to emergence, from rigidity to fluidity, and from isolated thinking to interconnected intelligence.

By embracing complexity, fostering continuous learning, and nurturing networked intelligence, you can build adaptive resilience. You will learn to grow stronger through disruption, seizing opportunities where others see obstacles—all while keeping the customer at the heart of your strategies.

The Non-Linear Imperative: Your Invitation to Explore

As you reflect on the stories of TechNova and GlobalSync, consider the critical choice facing your organization: Will you remain anchored in linear thinking, or will you embrace the dynamic, interconnected nature of true innovation focused on customer obsession? The pioneers of this approach are not merely changing processes—they are rewriting the narrative of how human creativity and technological potential intersect to meet and exceed customer expectations.

Are you ready to step beyond the straight line? The adventure of non-linear innovation is not about reaching a fixed destination; it's about continuously exploring an infinite landscape of possibilities. Together, we can reimagine what innovation can achieve in the realm of customer-obsessed strategic alliances.

Let your curiosity guide you as you discover how to harness the power of boundaryless customer engagement and unlock extraordinary potential in your strategic partnerships. Dive deep into this transformative journey and witness the remarkable impact that a non-linear approach can have on your organization and the customers you serve.

Summary: Embracing Customer Obsession

As we conclude this chapter on building a customer-obsessed strategic alliance, it is clear that placing the customer at the heart of every partnership is not merely a best practice but a necessity in today's competitive landscape. By focusing on aligning goals, maintaining open communication, and fostering a culture of innovation, organizations can create robust alliances that not only meet customer expectations but exceed them in delightful and unexpected ways.

Reflect on the insights shared in this chapter and consider how these principles can be woven into your own partnerships. Are you actively listening to your customers? Are you adapting your strategies based on their feedback? By continuously committing to a customer-centric approach, you can enhance collaboration and drive sustainable growth.

Remember, successful strategic alliances are built on trust, shared vision, and a relentless focus on delivering value. As you move forward, let this customer obsession guide your decisions and actions, ensuring that both your organization and your partners flourish in the ever-evolving market landscape.

Together, let's embrace the journey of creating partnerships that not only stand the test of time but also make a lasting impact on the lives of the customers we serve.

Practical Guidance:

Embrace Design Thinking for Boundaryless Customer Engagement

I invite you to embrace an innovative approach that can transform the way you engage with customers and partners: **design thinking**. This powerful methodology is not just a set of techniques; it is a mindset that prioritizes empathy, creativity, and collaboration.

By integrating design thinking into your non-linear initiatives and fostering a boundaryless approach to customer obsession, you can cultivate uncommon strategic partnerships that deliver exceptional value.

Practical Steps for Incorporating Design Thinking

In the ever-evolving landscape of business, organizations that wish to thrive must embrace innovative strategies that foster collaboration and prioritize customer needs. Incorporating design thinking into non-linear initiatives and boundaryless customer obsession is essential for cultivating uncommon strategic partnerships.

Below are practical steps that can guide organizations on this transformative journey.

1. Empathize with Customers and Strategic Partners

The first step in this process is to truly empathize with customers and partners. Start by conducting empathy interviews—schedule conversations with both groups to gather insights about their experiences, needs, and pain points. Use open-ended questions to encourage deeper discussions, enabling you to capture the essence of their thoughts and feelings.

Next, create customer journey maps to visualize the entire customer experience. By mapping out each touchpoint and interaction, you can identify areas ripe for improvement and innovation. Additionally, consider utilizing observational research; spend time observing customers in their natural environments to understand their behaviors and interactions with your products or services. This multifaceted approach lays the groundwork for deeper insights.

2. Define Core Challenges

Once you have gathered valuable insights, it's time to define the core challenges that need addressing. Organize collaborative workshops that bring together cross-functional teams to discuss the findings from your empathy activities. During these sessions, work together to articulate the core problems using "How might we…" statements, framing challenges in a way that inspires creativity and exploration.

Evaluate the identified challenges based on their potential impact on customer experience and feasibility. By prioritizing the most pressing issues, you set the stage for significant improvements that align with your customer obsession strategy.

3. Ideate Unconventionally

With clear challenges defined, it's time to unleash your creativity through ideation. Facilitate brainstorming sessions with diverse teams, encouraging all participants to contribute ideas without judgment. Utilize techniques like mind mapping or rapid ideation to explore as many ideas as possible.

Establish criteria to evaluate the generated ideas based on their ability to meet customer needs, align with organizational goals, and encourage collaboration among partners. This structured approach ensures that the most promising concepts are prioritized for further development.

4. Prototype Collaborative Solutions

Next, turn your best ideas into tangible prototypes. Focus on developing low-fidelity representations—these could be sketches, wireframes, or mock-ups—that allow for quick testing and iteration. Engage both internal stakeholders and customers in the prototyping process, gathering feedback early to identify areas for improvement and ensuring that your solutions align with customer expectations.

This hands-on approach not only fosters collaboration but also builds a sense of ownership among participants, reinforcing the value of their contributions.

5. Test and Iterate with Real Feedback

Once your prototypes are ready, it's time for user testing. Conduct sessions with actual customers to observe how they interact with your prototypes. Gather both qualitative and quantitative feedback to assess usability and satisfaction. This invaluable information serves as a compass for your next steps.

Encourage an iterative mindset within your teams, allowing for quick adjustments based on insights gained from testing. By remaining flexible and responsive, you can refine your solutions to better meet customer needs.

6. Create an Innovation Ecosystem

To sustain momentum, establish cross-functional teams that include members from different departments and partner organizations. This diversity fosters creativity and drives innovation. Encourage ongoing collaboration by organizing regular check-ins, brainstorming sessions, and innovation workshops. Utilize collaborative tools, such as digital whiteboards, to facilitate communication and idea-sharing among team members.

7. Embrace Boundaryless Collaboration

As you progress, host strategy workshops that involve all partners in the alliance. These sessions provide an opportunity to review progress, share insights, and refine strategies based on customer feedback and market dynamics. Building a culture of transparency is crucial—encourage open communication among partners by sharing successes, challenges, and insights. This transparency fosters trust and keeps all stakeholders aligned toward common goals.

8. Foster a Culture of Adaptability

Finally, empower employees at all levels to share ideas and propose solutions. Create an environment where experimentation is not just welcomed but encouraged. Recognize that failure is a learning opportunity and foster a mindset that views setbacks as stepping stones to success.

Provide training and resources that equip employees with design thinking skills. This could include workshops, online courses, or access to innovation toolkits, all designed to enhance their ability to think creatively and collaboratively.

Conclusion

By following these practical steps, organizations can effectively integrate design thinking into their non-linear initiatives and boundaryless customer obsession. This approach not only enhances collaboration with strategic partners but also fosters innovation that resonates deeply with customers. The ultimate goal is to create meaningful value that strengthens customer relationships and drives sustainable growth. Together, let's embrace this transformative journey and unlock the potential of uncommon strategic partnerships!

As you embark on this exciting path, remember that the power of collaboration, empathy, and creativity can redefine your approach to customer engagement, leading to unparalleled success in today's dynamic market landscape.

The future is bright, and the possibilities are endless, in words of Nike, "Just do it." Let's move to chapter 4 where I will introduce the concept of The Equilateral Triangle, which I consider to be the cornerstone of customer-obsessed strategic alliance.

"Collaboration allows us to know more than we are capable of knowing by ourselves." — Paul Solarz

Chapter 4

The Equilateral Triangle

Introduction: The Apex of Customer Centricity

In today's ever-evolving business landscape, customer centricity is not merely a catchphrase; it stands as the cornerstone of successful strategic alliances. As you strive to innovate and maintain a competitive edge, it is crucial to prioritize your management operating model around the needs of your customers.

Visualize an equilateral triangle where customer centricity reigns supreme at the apex, shining like a guiding star. At the base of this triangle, we find two vital players: Company A, a leading Cloud Service Provider (CSP) such as AWS, Azure, or GCP, and Company B, a Global Service Integrator (GSI) renowned for its expertise in leveraging technology to transform businesses. This dynamic illustrates how aligning the efforts of both companies around a shared commitment to the customer can lead to extraordinary outcomes.

The Apex: The Customer

At the pinnacle of our triangle sits the customer, representing the ultimate objective of any strategic alliance. In this context, customers are not just passive recipients of products or services; they are active participants whose preferences and feedback play a pivotal role in shaping the direction of both organizations. In a market flooded with choices, customer loyalty hinges on your ability to understand and genuinely meet their needs. **This understanding is paramount for any organization aspiring to progress.**

Consider Company A as it prepares to launch a new cloud-based platform. Initially, their focus might be squarely on the technology's features and capabilities. However, without a profound understanding of what customers truly value, ease of use, seamless integration with existing systems, or cost-effectiveness, they risk creating a product that misses the mark in the marketplace.

By placing the customer at the heart of their strategic discussions, Company A can gather insights that inform the platform's development, ensuring it addresses genuine pain points. In this paradigm, the customer is not merely a recipient of services; they are the heartbeat of the partnership, propelling the strategies and innovations that define the collaboration.

Aligning Company A and Company B

Now, let's turn our attention to Company B, the Global Service Integrator. This organization brings invaluable expertise in deciphering customer needs and designing tailored solutions that maximize the benefits of the cloud technology offered by Company A. When both companies align their strategies around customer obsession, they create a potent synergy.

Communication plays a vital role in this partnership. For instance, Company B can share its insights gathered from working with clients across various industries, offering invaluable feedback that informs Company A's product features. This iterative process enables both organizations to refine their offerings, ultimately elevating the customer experience. Imagine a scenario where Company B conducts interviews with customers to uncover how businesses utilize cloud technology. They discover that many clients struggle with integrating cloud services into their existing workflows. Armed with this knowledge, both companies can collaborate to develop user-friendly integration tools or tailor training sessions that empower clients to maximize their cloud investments. Recognizing and addressing customer needs marks a critical recovery of your organizational priorities; true customer centricity always triumphs in the long run.

Climbing Higher Together

As Company A and Company B continue to collaborate and innovate, they draw closer to each other, ascending toward the apex of the triangle—the customer. Their shared commitment to understanding and addressing customer needs creates a product that not only meets but exceeds expectations.

This journey is not without its challenges. Misalignment in goals or communication can lead to setbacks, yet by keeping customer obsession at the forefront, both companies are better equipped to navigate these hurdles. They learn from feedback, adjust their strategies, and remain flexible in response to changing market dynamics. For instance, if initial customer feedback reveals that certain features of the cloud platform fall short, both organizations can pivot quickly to address these concerns, demonstrating their unwavering dedication to delivering value.

The Peak: A Superior Customer Experience

When Company A and Company B successfully align their efforts around customer obsession, they craft an exceptional customer experience that distinguishes them in the marketplace. At this pinnacle of the triangle, customers feel valued and engaged, perceiving both companies as partners genuinely invested in their success rather than mere service providers. This heightened focus on customer needs fosters brand loyalty, transforming customers into advocates. Satisfied customers are likely to share their positive experiences with others, generating referrals and enhancing the reputations of both organizations. The result is a virtuous cycle where collaboration fuels innovation, leading to more satisfied customers and greater success for the partnership.

Embracing Customer Obsession

Placing the customer at the apex of the equilateral triangle is not just a strategy; it represents a fundamental shift in how you approach your alliances. By fostering an environment of customer obsession, Company A and Company B can collaborate effectively, leveraging their strengths to create solutions that resonate deeply with their clients.

As you navigate your strategic alliances, make it a priority to elevate the customer's voice and experience. Doing so will enable you to ascend higher in the triangle, achieving remarkable results and laying the groundwork for sustained success. Together, you can unlock extraordinary possibilities, ensuring that both the partnership and the customers flourish. Something remarkable occurs when teams come together: they create a synergy that enhances their understanding of customer needs.

True potential can only be realized when organizations unite with a shared vision centered around the customer. This elevated focus on customer obsession not only enhances brand loyalty but also drives mutual success. It is at this stage of strategic partnership that the customer becomes your advocate! This is where your customer shares their positive experiences with colleagues and peers, creating a ripple effect that benefits both companies. The voice of the customer generates valuable insights and feedback that continue to fuel future innovations, keeping the partnership agile and responsive.

Reflecting on Your Organization

Now, let's turn our attention to your organization. How well do you understand your customers? Here are some reflective questions to consider:

- What processes do you currently have in place to gather and act on customer feedback? Are you utilizing a mix of quantitative and qualitative methods to gain a complete picture of customer sentiments?
- How effectively do you integrate customer insights into your decision-making processes? Are there structured systems that allow feedback to flow seamlessly into product development and service enhancements?

- What additional methods could you implement to deepen this understanding? Could you host customer advisory boards or create community forums where users can voice their opinions and suggestions?

Taking time to ponder these questions can help you uncover new opportunities to connect with your customers on a deeper level and enhance your offerings in meaningful ways. Understanding customer needs is not a one-time event but an ongoing journey that requires commitment and adaptability from everyone in the organization. Embracing this mindset will pave the way for lasting customer relationships and drive the success of your strategic alliances.

The Importance of Agility in Execution

In today's fast-paced world, customer preferences can shift overnight. Therefore, you must be prepared to respond swiftly. This agility requires open communication channels, cross-functional collaboration, and a willingness to experiment. Companies that can pivot quickly in response to customer feedback are more likely to maintain their competitive edge.

Consider the story of a popular food delivery service that faced intense competition in its market. As they began to gather feedback from customers, they noticed a rising demand for healthier meal options. Rather than sticking to their traditional offerings, the leadership team quickly rallied to develop partnerships with local organic farms and health-conscious restaurants.

This agile approach allowed them to adapt to their menu in real-time, offering fresh, nutritious meal kits that not only satisfied customer cravings but also attracted a new customer base eager for healthier choices. By being responsive and attentive to their customers' evolving needs, they transformed potential stagnation into growth.

Building Trust Through Execution

At its core, customer obsession means prioritizing the needs and desires of customers above all else. This involves cultivating a culture where the voice of the customer is integral to decision-making processes, from product development to service delivery.

When organizations embrace this philosophy, they do more than just meet customer expectations; they exceed them, creating a loyal base that can propel growth and innovation.

The Impact of Commitment on Customer Relationships

A commitment to the partnership translates into a commitment to your customers. Engaged partners are more likely to stay attuned to customer feedback, allowing for quick adjustments to strategies based on what customers are saying.

When organizations consistently deliver on their promises, they demonstrate their commitment to serving their customers well, fostering a sense of trust and appreciation.

Picture a pizza restaurant that prides itself on being the fastest delivery service in town. One day, they received an influx of feedback from customers complaining about the quality of their food. Instead of blaming their suppliers or making excuses, the restaurant decided to go all-in on quality. They held a meeting, ordered some gourmet pizza for inspiration, and devised a plan to elevate their ingredients and recipes. Now, they may not be the fastest, but they've certainly become the most beloved!

This delightful turnaround shows how listening to customers can lead to significant improvements. Are you ready to embrace this journey of growth and partnership?

Here are five key lessons learned from this chapter on the importance of customer obsession with strategic alliances:

1. **Customer Centricity as the Cornerstone:** Placing the customer at the apex of the strategic alliance triangle is essential. This mindset transforms customers from passive recipients into active participants whose feedback and needs shape the collaboration. Companies that prioritize customer understanding can better align their strategies, resulting in stronger partnerships and enhanced customer satisfaction.

2. **Collaborative Synergy Enhances Value:** Effective communication and collaboration between partners are vital. When companies like Company A (CSP) and Company B (GSI) share insights and actively engage in iterative processes, they can refine their offerings based on real customer feedback. This synergy not only improves product development but also creates tailored solutions that maximize customer value.

3. **Agility in Execution Drives Success:** In a rapidly changing marketplace, organizations must be agile and responsive to customer feedback. Maintaining open communication channels and fostering a culture that embraces flexibility enables partners to pivot quickly and adapt their strategies to meet evolving customer needs. This adaptability is key to maintaining a competitive edge.

4. **Continuous Learning and Improvement:** Understanding customer needs is an ongoing journey. Companies should establish processes for gathering and integrating customer feedback into decision-making continuously. Regularly assessing customer sentiments and adapting based on insights will help organizations remain relevant and ensure long-term success.

5. **Building Trust through Commitment:** A commitment to the partnership is intrinsically linked to a commitment to the customer. When organizations consistently deliver on their promises and prioritize customer needs, they foster trust and loyalty. This dedication leads to stronger relationships, where satisfied customers become advocates, enhancing brand reputation and driving further growth.

These lessons collectively emphasize that true customer obsession is a strategic imperative that not only strengthens partnerships but also elevates the overall customer experience. As we conclude this chapter, it's clear that embedding customer obsession into the fabric of strategic alliances is not just a best practice, it's a transformative approach that can propel organizations to new heights.

By prioritizing customer needs and fostering collaborative synergy, companies can create lasting value that resonates deeply with their clients. Now, as we turn the page to Chapter 5: Measuring Customer Success — The Right KPIs and Metrics, we will delve into the crucial topic of how to quantify that success effectively.

Understanding which key performance indicators (KPIs) and metrics truly reflect customer satisfaction and engagement is vital for sustaining growth and refining strategies. After all, what gets measured gets managed, and the right metrics will illuminate the path to continuous improvement and long-term success.

Let us continue our exploration of how to harness data to drive meaningful insights, ensuring that your customer-centric efforts translate into tangible results!

Chapter 5

Measuring Customer Success — The Right KPIs & Metrics

In this pivotal section of our journey, I'll explore the essential topic of measuring customer success. Understanding how well your customers are realizing business value is critical for any organization striving for excellence. But how do you determine if your customers are truly thriving?

Here, I'll introduce you to customer-centric key performance indicators (KPIs) and associated metrics that will serve as your guiding compass. These tools will not only help you gauge the effectiveness of your products and services but also empower you to make informed decisions that enhance customer satisfaction and loyalty.

By the end of this chapter, you'll be equipped with the knowledge to track success effectively, ensuring that your partnerships continue to deliver meaningful value to your customers. Let's embark on this journey to unlock the true potential of your customer relationships!

In the world of strategic alliances, keeping the customer at the heart of everything is essential. By focusing on customer-centric KPIs, companies can enhance their offerings and strengthen their partnerships. So, let's delve into the metrics that matter most and illustrate their impact using the Equilateral Triangle analogy, featuring Company A, a Cloud Service Provider, and Company B, a Global Service Integrator. Together, they can achieve success by working closely with their customers and leveraging insights from their collaborative efforts.

Customer Satisfaction Score (CSAT)

One of the first metrics we'll discuss is the Customer Satisfaction Score (CSAT). This crucial KPI measures how satisfied customers are with the products or services provided by the alliance. Typically, it's assessed through surveys where customers rate their satisfaction on a scale (for example, from 1 to 5). Keeping the questions simple and direct allows you to gather clear insights, and follow-up questions can help you understand why customers rated you the way they did.

Imagine Company A has just launched a new cloud-based platform in partnership with Company B to provide seamless integration for clients. After the launch, they send out a CSAT survey, and customers respond with an impressive average rating of 4.5. However, feedback reveals that the onboarding process was a bit confusing. Recognizing this, both companies can collaborate to enhance the onboarding experience, ensuring it's user-friendly and intuitive. This commitment to customer feedback leads to higher satisfaction with future releases.

Net Promoter Score (NPS)

Next, let's explore the Net Promoter Score (NPS), which gauges customer loyalty and their likelihood of recommending the service to others. Customers are asked to rate their likelihood of recommendation on a scale from 0 to 10. A score of 9 or 10 indicates a promoter, while scores from 0 to 6 suggest a detractor. Analyzing comments helps organizations understand what drives loyalty or dissatisfaction.

After a successful product launch, Company A and Company B conducted an NPS survey. To their delight, 80% of customers were promoters, citing the smooth integration as a key factor. However, a few detractors expressed concerns about pricing. Armed with this knowledge, the two companies can explore pricing strategies that retain customer loyalty while addressing the concerns of those who are hesitant to recommend the product.

Customer Retention Rate

The Customer Retention Rate is another vital metric that measures the percentage of customers who continue to use the product or service over a specified time. This rate is calculated by taking the number of customers at the end of the period, subtracting new customers acquired during that period, and dividing them by the number of customers at the start of the period. A high retention rate indicates customer satisfaction and loyalty.

After the first year of collaboration, Company A and Company B track their retention rates and find that 90% of their clients continue using the cloud platform. While they celebrate this success, they also investigate the reasons behind the 10% who chose to leave. Through discussions, they discover that some clients felt unsupported during their transition. This insight prompts them to enhance their customer support services, further improving retention rates in the future.

Customer Lifetime Value (CLV)

Customer Lifetime Value (CLV) estimates the total revenue a customer is expected to generate throughout their relationship with the alliance. CLV is calculated by multiplying the average purchase value by the number of transactions per year and then by the average customer lifespan. Understanding CLV helps partners identify how much they can invest in customer acquisition and retention strategies.

Upon analyzing their CLV, Company A and Company B realize that customers using both the cloud service and integration solutions generate significantly higher revenue over time. They have decided to implement targeted marketing campaigns aimed at onboarding new clients to both services, ensuring they maximize their CLV and foster deeper relationships with these customers.

Customer Feedback and Insights

Collecting qualitative and quantitative feedback on the alliance's products or services is vital. Regularly gathering insights through surveys, focus groups, and direct interviews provides actionable information that can guide product improvements and strategy adjustments.

After launching their new services, Company A and Company B host a series of focus groups to gather customer feedback. Clients share their experiences and suggest additional features that could enhance functionality. By actively listening and incorporating this feedback, the companies strengthen their offerings and demonstrate to customers that their opinions matter.

Churn Rate

The Churn Rate measures the percentage of customers who stop using the product or service within a given timeframe. This is calculated by taking the number of customers lost during a period, dividing it by the total number of customers at the start of that period, and then multiplying by 100. A high churn rate can indicate dissatisfaction or unmet needs.

Company A and Company B notice an uptick in their churn rate after a major product update. Concerned, they conduct exit interviews with departing customers and learn that the new interface is confusing for many. Armed with this information, they work together to simplify the user interface and provide additional training for existing customers, helping to reduce future churn.

First Contact Resolution Rate

Another important metric is the First Contact Resolution Rate, which measures the percentage of customer issues resolved on the first interaction. This is calculated by dividing the number of issues resolved on first contact by the total number of issues. A high-resolution rate indicates effective support processes.

Company A and Company B implement a joint customer support initiative and monitor their First Contact Resolution Rate. They discover that they're resolving

85% of inquiries on the first call. To further improve this, they create a shared knowledge base accessible to both support teams, allowing for quicker resolutions and enhanced overall customer experience.

Time to Market for New Offerings

The Time to Market for New Offerings measures the time taken to develop and launch new products or services resulting from the alliance. This is assessed by evaluating the time from project initiation to market launch. A faster time to market can provide a competitive advantage.

Recognizing that the time to market for new features was slowing down their competitiveness, Company A and Company B held a brainstorming session. They streamline their development process and improve communication, resulting in a 30% reduction in launch time for new features. Customers benefit from quicker access to innovative tools, reinforcing the partnership's value.

Usage Metrics

Usage Metrics track how frequently customers use the products or services provided by the alliance. This involves analyzing usage data, such as the number of logins, transaction volumes, or feature engagement. High usage rates can indicate customer satisfaction and product value.

Company A and Company B notice that a certain feature of their platform is rarely used by customers. Curious, they investigate and discover that customers aren't aware of its benefits. They launch a campaign to promote this feature, complete with tutorials and case studies showcasing its advantages. The result is increased usage and satisfied customers who feel empowered by the enhanced functionality.

Customer Acquisition Cost (CAC)

Finally, Customer Acquisition Cost (CAC) measures the cost associated with acquiring a new customer through the partnership. This is calculated by dividing total marketing and sales expenses by the number of new customers acquired in each period. Understanding CAC helps inform marketing strategies and budgeting.

After tracking their CAC, Company A and Company B find it higher than anticipated. They analyze their marketing spend and discover that some campaigns aren't delivering results. By collaborating on a targeted marketing approach that leverages both companies' strengths, they significantly reduce their CAC while attracting high-quality leads.

Concluding thoughts:

By focusing on these customer-centric KPIs, Company A and Company B can ensure they are meeting customer needs effectively and building lasting relationships. Tracking these metrics allows organizations to adapt their strategies based on real customer feedback, ultimately driving growth and success for both partners.

Prioritizing the customer in this way not only enhances satisfaction and loyalty but also strengthens the overall partnership, leading to a win-win situation for all involved.

As we continue our journey, we'll delve into how to measure the success of these partner-centric KPIs in the next chapter. Get ready to explore the tools and techniques that will help you gauge the effectiveness of your strategic alliances and ensure that both your organization and your customers flourish together!

Chapter 6

Strategic Alliance Centric Key Performance Indicators (KPIs) – Measuring Growth and Success

As we delve deeper into the realm of strategic alliances, two crucial aspects stand out: governance and measuring success. Just like a flourishing garden requires regular care and attention, your partnership needs ongoing monitoring and assessment to thrive. When it comes to strategic alliances, understanding whether both parties are benefiting is essential for long-term growth and sustainability.

The Importance of Governance in Measuring Key Metrics

As you navigate the landscape of strategic alliances, it's crucial to recognize that effective governance is not just a best practice—it's the cornerstone of your success. Establishing a robust governance framework enables your organization to measure key metrics accurately and optimize collaborative efforts with your partners. By prioritizing customer-centric objectives within this framework, you can create a foundation that supports sustained success in your alliances.

Customer-Centric Governance

At the heart of effective governance lies customer-centricity. You should actively engage key stakeholders from both your organization and your partners early in the process to ensure that all voices—including those of your customers—are heard and considered. This proactive approach allows you to create joint governance structures that respect the interests of everyone involved, facilitating the development of shared vision statements that resonate across all stakeholders. Implementing cross-organizational feedback mechanisms will enable your team to continuously adapt strategies based on real customer insights.

Enhancing Collaboration for Customer Success

Successful strategic alliances thrive on collaboration, and your organization must embrace this principle. Open communication and knowledge sharing are vital. When you and your partners prioritize customer needs and work together to refine your offerings, you create tailored solutions that maximize value for your clients. This synergy not only strengthens the relationship between your organizations but also leads to better products and services for the customers you serve.

Agility and Responsiveness

In today's fast-paced business environment, agility is essential. You need governance structures that facilitate realistic partnership planning and effective management of resistance. This will allow your organization to respond swiftly to changes in customer preferences. As you adapt to evolving customer needs, having a flexible framework ensures that both you and your partners can pivot quickly, maintaining alignment with your ultimate goal: delivering exceptional value to your clients.

Continuous Improvement and Learning

A commitment to continuous improvement and learning is vital for customer-centric strategic alliances. By implementing joint learning and review processes, you can systematically capture insights and adapt your strategies based on feedback. This approach aligns perfectly with the theme of your journey toward customer obsession, as it emphasizes using customer insights to drive innovation and enhance partnerships. Organizations that embrace this philosophy will be better positioned to refine their strategies and create more meaningful customer experiences.

Long-Term Relationship Building

Finally, focus on sustainable partnership development emphasizes the necessity of building long-term relationships in strategic alliances. By prioritizing governance that supports enduring collaboration, you can cultivate trust and loyalty—not just between your organization and your partners, but also with your customers. This creates a virtuous cycle where satisfied customers become advocates, further driving your business success.

In essence, the principles of effective governance outlined here are essential for cultivating customer-obsessed strategic alliances. They provide a structured approach that emphasizes collaboration, agility, continuous improvement, and long-term relationship building—key elements for delivering exceptional customer value and driving sustainable growth. Now let us turn our attention to measuring key performance indicators.

Measuring Key Performance Indicators

It is essential to focus on key performance indicators (KPIs) that serve as guiding stars, helping you evaluate and enhance your collaborative efforts. By concentrating on customer-centric KPIs, you will gain valuable insights that not only track the effectiveness of your alliances but also empower you to make informed decisions that benefit everyone involved.

Let's dive into some recommended KPIs, supported by real-world examples that highlight their impact and significance.

Revenue Growth

One of the most straightforward yet critical KPIs is revenue growth. This metric acts as a financial health check-up for your partnership, tracking the increase in revenue generated directly from the alliance. If both organizations work well together, you should witness a positive trend in revenue—much like a thriving plant flourishing in sunlight.

Key Points to Consider:

- **Direct Attribution:** Ensure that revenue increases can be directly tied to the alliance by analyzing sales data before and after the partnership began. This is akin to tracking the progress of a diet—you need to know where you started to appreciate how far you've come!
- **Setting Targets:** Establish clear revenue targets based on realistic forecasts to guide collaborative efforts and keep everyone focused on the goal.
- **Regular Reviews:** Regularly check in on revenue growth. If you notice stagnation or decline, it might be time to reassess your strategies or identify new opportunities together.

Consider the example of Company A, a Cloud Service Provider, and Company B, a Global Service Integrator. After forming their alliance, they set a revenue growth target of 20% in the first year. By closely monitoring sales data and attributing increases to their joint marketing campaigns, they achieved an impressive 30% increase in revenue!

This success motivated both companies to continue innovating together, setting even higher targets for the next year. It's like having a motivational workout buddy pushing you to reach new heights!

Cost Savings

Next on the list is cost savings, a crucial metric that assesses reductions in operational costs achieved through collaboration. This KPI is akin to finding a way to do more with less—always a win-win!

Key Points to Consider:

- **Identifying Areas for Savings:** Look for specific areas where costs can be cut through shared resources, such as joint marketing efforts or combined purchasing power. It's like pooling your resources for a group gift; it's often cheaper and much more enjoyable!
- **Measuring Impact:** Document changes in expenses provide a clear picture of how the alliance contributes to cost savings.
- **Continual Improvement:** Foster ongoing discussions about cost efficiencies, as new opportunities for savings may arise as both companies grow.

In their partnership, Company A and Company B identified that by pooling their marketing resources, they could significantly cut costs. They jointly developed a campaign that saved them 25% on marketing expenses while also improving their outreach.

The success of this initiative prompted them to explore further areas for collaboration, such as shared logistics or joint training programs. It's like a potluck dinner where everyone brings something to the table—together, they create a feast!

Market Share Growth

Monitoring market share growth is another critical KPI. This metric measures changes in market share due to the alliance, reflecting the partnership's competitive impact. It's like checking how well your team is performing against others in the league!

Key Points to Consider:

- **Data Sources:** Use reliable market data to track combined market share over time, utilizing industry reports or market research surveys.
- **Competitor Analysis:** Keep an eye on competitors to understand how your alliance positions you in the market. Are you gaining ground, or do you need to adjust your strategies?
- **Customer Insights:** Gather feedback from customers to understand their perception of your alliance and its products.

Company A and Company B launched a new joint offering targeting a previously underserved market segment. After six months, they analyzed market share data and discovered a 15% increase in their combined share, significantly outpacing their closest competitor. This success reinforced their commitment to collaborative innovation and encouraged them to explore additional market opportunities together, proving that teamwork indeed makes the dream work!

Customer Satisfaction

Customer satisfaction is a vital KPI that evaluates feedback and satisfaction levels for jointly developed products or services. Happy customers are often the best indicators of a successful partnership!

Key Points to Consider:

- **Feedback Mechanisms:** Implement surveys, focus groups, or customer interviews to gather insights into their experiences with your joint offerings. Think of this as a heartfelt chat with a friend; you want to know how they truly feel!
- **Actionable Insights:** Look for trends in customer feedback. If customers express concerns, address them collaboratively to improve satisfaction.
- **Continuous Engagement:** Keep communication lines open with your customers. Their evolving needs can guide your alliance's future strategies.

After rolling out their new cloud platform, Company A and Company B conducted a series of customer satisfaction surveys. They discovered that while most customers were pleased with the integration, some struggled with specific features. In response, they quickly organized a joint customer support initiative to address these concerns. By implementing changes based on customer feedback, they significantly improved satisfaction scores, demonstrating their commitment to the customer experience. It's like a dance; when both partners are in sync, it's a joy to behold!

Innovation Outcomes

Tracking innovative outcomes involves measuring the number of new products or services created through the alliance and assessing their market performance. This metric highlights the creative potential of your collaboration.

Key Points to Consider:

- **Tracking New Developments:** Keep a log of new innovations resulting from your partnership. This can include everything from new features to entirely new product lines.

- **Market Impact:** Analyze how these innovations perform in the market. Are they meeting customer needs? Are sales reflecting their success?

- **Fostering a Culture of Innovation:** Encourage brainstorming sessions and creative thinking among your teams. The more ideas are shared, the more opportunities for innovative solutions.

Company A and Company B decided to host quarterly innovation workshops where teams from both companies could brainstorm new features for their cloud platform. After the first workshop, they developed a groundbreaking data analytics tool that customers raved about. The success of this tool not only enhanced their product offerings but also solidified their reputation as industry leaders in innovation. It's a classic case of "two heads are better than one!"

Engagement Metrics

Last but certainly not least, we have engagement metrics. This KPI measures the frequency and quality of interactions between partners, helping you assess the health of the relationship. Think of it as a relationship check-in—how connected are you and your partner?

Key Points to Consider:

- **Communication Frequency:** Track how often you meet, share updates, or collaborate on projects. Regular touchpoints are crucial for maintaining a strong partnership.
- **Quality of Interactions:** Evaluate the effectiveness of these interactions. Are discussions productive? Are both parties actively contributing and engaging?
- **Feedback Loops:** Establish ways for both teams to provide feedback on partnership dynamics. This can help identify areas for improvement in engagement.

Company A and Company B implemented a regular meeting schedule to ensure they stayed connected. During these meetings, they not only discussed project updates but also shared insights on industry trends. They found that these collaborative discussions enhanced both the quality of their work and their relationship. By fostering open lines of communication, they created a partnership where both teams felt empowered and engaged, proving that a little communication goes a long way!

The Importance of Focusing on KPIs

By focusing on these relevant KPIs, your organization can gain critical insights into performance, identify areas for improvement, and make informed decisions that drive growth.

- **Ensuring Alignment with Objectives:** KPIs help ensure that your partnership remains aligned with strategic objectives. Regularly measuring relevant metrics confirms that both parties are working toward shared goals.
- **Driving Accountability:** Implementing KPIs fosters a culture of accountability within partnerships. When both parties know their performance will be evaluated against specific metrics, they are more likely to stay committed and engaged in the collaboration.
- **Enhancing Customer Satisfaction:** Customer-centric KPIs are vital for understanding how well your alliance meets customer needs. Tracking metrics like Customer Satisfaction Score (CSAT) or Net Promoter Score (NPS) helps gauge the effectiveness of your joint offerings.
- **Identifying Opportunities for Improvement:** Regularly measuring KPIs allows your organization to analyze performance trends over time. This analysis reveals strengths and weaknesses in the partnership, providing a roadmap for continuous improvement.
- **Supporting Strategic Growth:** KPIs provide insights that inform long-term growth strategies. Understanding how your alliance performs against established metrics enables your organization to make strategic adjustments that enhance market position.

In summary, measuring KPIs in strategic alliances is not just a best practice; it is essential for ensuring the partnership's success and sustainability. By implementing a robust KPI framework, you can ensure alignment with objectives, drive accountability, enhance customer satisfaction, identify improvement opportunities, and support strategic growth. Neglecting to measure these indicators can lead to misalignment, disengagement, customer dissatisfaction, stagnation, and lost opportunities. In a competitive landscape, understanding and utilizing KPIs can make the difference between a thriving partnership and one that falters.

By keeping a close eye on these partner-centric KPIs, your organization can ensure your strategic alliance is on the right track. Each metric provides valuable insights into different aspects of the partnership, helping you make informed decisions and adjustments along the way. So, take the time to measure and reflect on these key performance indicators, and watch your alliances flourish!

As we transition to the next chapter, prepare to explore the tools and techniques that will help you gauge the effectiveness of your partnerships and ensure that both your organization and your customers thrive together!

Self-Reflection Questions

To further engage with the concepts discussed in this chapter, consider these self-reflection questions:

1. How effectively are we measuring revenue growth in our partnerships?
2. What steps can we take to enhance our customer satisfaction metrics?
3. Are we actively analyzing our market share and how it reflects our alliance's success?
4. How do we foster a culture of innovation between our teams?
5. Are our engagement metrics sufficient to gauge the health of our partnership?
6. What actions can we take to ensure continuous improvement in cost savings?
7. How often do we check in on our key performance indicators?
8. Are we aligning our KPIs with our strategic objectives effectively?
9. How do we incorporate customer feedback into our decision-making processes?
10. What are the most significant areas for improvement in our current measurement strategies?

Take a moment to ponder these questions and see how they can help refine your approach to measuring success in your strategic alliances!

With a commitment to collaboration and a focus on customer needs, the sky's the limit for what these partnerships can achieve!

Chapter 7

Collaboration and Innovation in Strategic Alliances

In today's fast-paced business environment, collaboration and innovation have become the twin engines driving successful strategic alliances. As organizations navigate the complexities of the modern marketplace, they must learn to work together in ways that foster creativity and adaptability. This chapter explores the importance of these elements, drawing insights from influential works that illuminate the path to thriving partnerships.

The Collaborative Mindset

To understand the essence of collaboration, we must first recognize that it goes beyond merely working together; it embodies a spirit of synergy. In *Collaborative Advantage: How Collaboration Beats Competition as a Strategy for Success*, Charles Handy articulates that collaboration produces outcomes greater than the sum of individual contributions. This mindset is crucial for strategic alliances, where each partner brings unique strengths to the table.

Think of it this way: when organizations collaborate effectively, they can achieve extraordinary results that neither could accomplish alone. Picture a tech startup that specializes in innovative software solutions partnering with a well-established manufacturing firm. The startup provides cutting-edge software, while the manufacturer contributes years of industry experience. Together, they can create groundbreaking products that not only meet but exceed customer expectations.

This collaborative approach requires more than just bringing different teams together; it demands open communication, mutual respect, and a shared vision. Partners need to understand their strengths and how to leverage them effectively. This creates an environment where all voices are heard and valued, encouraging team members to share their ideas freely.

Imagine two companies - one a tech startup focused on developing smart appliances and the other a manufacturing giant with decades of expertise. By joining forces, the startup can integrate innovative software that learns from user behavior, while the manufacturer ensures top-notch production quality. The result?

A product that not only meets customer needs but anticipates them, creating a delightful user experience. This synergy illustrates how collaboration can lead to innovations that set a new standard in the market.

Lessons Learned:

1. **Diverse Perspectives Matter:** When organizations collaborate effectively, they can achieve extraordinary results that neither could accomplish alone. I have seen firsthand how diverse viewpoints can spark innovative ideas that lead to successful outcomes.

2. **Open Communication is Key:** Creating an environment of open communication fosters an atmosphere where all voices are heard, encouraging team members to share their ideas freely.

3. **Shared Vision Drives Success:** Developing a shared vision aligns all parties toward common goals, ensuring everyone is invested in the partnership's success.

Imagine the remarkable potential when organizations collaborate effectively. Picture a tech startup specializing in innovative software solutions joining forces with a well-established manufacturing firm. The startup brings cutting-edge technology, while the manufacturer contributes decades of industry expertise. Together, they have the ability to create groundbreaking products that not only meet but exceed customer expectations.

Reflective Challenge: How can you encourage a culture where all voices are heard and valued within your organization? What specific steps can you take to foster an environment that nurtures collaboration among your team members and partners?

Innovation as a Necessity

As we shift our focus to innovation, we find it is a crucial component of successful alliances. In *The Innovator's Dilemma: When New Technologies Cause Great Firms to Fail*, Clayton M. Christensen emphasizes how established companies often struggle to innovate because they focus too heavily on existing customers and technologies. This is where collaboration can be a game-changer. By teaming up with partners who offer different perspectives and expertise, organizations can foster a culture of innovation that challenges the status quo.

The reality is that the business landscape is rife with stories of companies that failed to adapt to disruptive innovations, often because they became too comfortable in their established ways. Collaboration allows firms to share insights and create an environment that encourages experimentation. When organizations join forces, they can explore unconventional ideas and strategies that might not have been possible if they operated in isolation.

Consider the collaboration between Spotify and Uber, which allowed Uber riders to control their music during trips. This innovative feature was born from the mutual understanding that both companies wanted to enhance the customer experience. By pooling their resources and insights, they created something truly unique that delighted users and distinguished them from the competition.

Moreover, Eric Ries, in *The Lean Startup*, underscores the importance of continuous innovation through experimentation and learning. Organizations must not only be open to new ideas but also be willing to test and iterate. Strategic alliances provide fertile ground for this experimentation, allowing partners to innovate together and adapt quickly to market changes.

Imagine a software company that recognizes the need to improve its user interface based on customer feedback. Instead of adhering to a rigid development schedule, the company adopts a flexible process that enables real-time adjustments based on user input. This agile approach leads to enhanced customer satisfaction and positions the company as a leader in the industry, demonstrating how collaboration and commitment to innovation can yield significant competitive advantages.

Lessons Learned:

1. **Experimentation is Essential:** Being open to experimenting with new ideas and approaches is vital for fostering innovation. I have seen how organizations thrive when they embrace a mindset of continuous improvement.
2. **Learning from Failure:** Embracing failures as learning opportunities has taught me that each setback can pave the way for future successes.
3. **Cross-Industry Collaborations Spark Ideas:** Collaborating with partners from different industries often yields fresh insights and innovative solutions.

The reality is that the business landscape is filled with stories of companies that failed to adapt to disruptive innovations, often because they became too comfortable with established ways of doing things.

When you and your partners join forces, you can share insights and create an environment that encourages experimentation.

Consider the collaboration between Spotify and Uber, which enabled Uber riders to control their music during trips. This innovative feature was born from a mutual understanding of enhancing customer experiences. Are you actively seeking opportunities to innovate alongside your partners? How can you create an environment that fosters creativity and encourages unconventional ideas?

Reflective Challenge: What innovations can your organization pursue that leverage your existing partnerships? Are there unexplored ideas or strategies that could benefit from a collaborative approach?

Customer-Centric Collaboration

A vital aspect of collaboration and innovation is the focus on customer experience. In *Customer Experience 3.0: High-Profit Strategies in the New Economy*, John A. Goodman discusses how businesses that prioritize the customer experience are more likely to succeed. Strategic alliances should focus on understanding and enhancing this experience, making it essential for partners to share insights and feedback from customers.

Integrating customer feedback should be a joint effort. When both partners actively seek and incorporate this feedback into their strategies, they create a more comprehensive understanding of their customers' desires. This collaboration not only strengthens the partnership but also ensures that the final product truly resonates with the target audience.

Imagine a partnership between a retail chain and a logistics company that closely collaborates to develop an innovative supply chain solution. This partnership enables faster delivery times and real-time tracking, significantly enhancing the customer experience. By pooling their knowledge and resources, they create a seamless process that directly addresses customer pain points, demonstrating the power of collaboration.

Lessons Learned:

1. **Customer Feedback is Gold:** Integrating customer feedback into the collaboration process is crucial for success. I have found that proactive outreach can yield invaluable insights.
2. **Joint Efforts Yield Better Results:** Collaborating to seek customer insights strengthens the partnership and enhances the final product.
3. **Proactive Problem-Solving:** Addressing customer pain points together creates a more effective and appealing offering.

Integrating customer feedback should be a joint effort. When you and your partners actively seek and incorporate this feedback into your strategies, you create a more comprehensive understanding of your customers' needs.

Imagine a partnership between a retail chain and a logistics company closely collaborating to develop an innovative supply chain solution. This partnership enhances the customer experience through faster delivery times and real-time tracking. How can you leverage your partnerships to address customer pain points effectively?

Reflective Challenge: How is your organization currently integrating customer feedback into its strategic alliances? Are there opportunities for deeper collaboration that could enhance the overall customer experience?

Taking a moment to ponder these questions can lead to actionable insights that strengthen your partnership.

Trust and Relationship Building

Collaboration thrives in an environment of trust. As discussed in *The Trusted Advisor* by David H. Meister, Charles H. Green, and Robert M. Galford, trust is the foundation of any successful partnership. When partners trust one another, they are more likely to share ideas freely, take risks, and collaboratively navigate challenges.

Building trust requires consistent communication and transparency. Partners should openly discuss their goals, concerns, and expectations from the outset. This openness lays the groundwork for a strong relationship and fosters a sense of partnership.

Imagine a marketing agency collaborating with a client to launch a new campaign. When both parties trust each other, the agency can suggest bold, innovative ideas without fear of overstepping boundaries. This open dialogue nurtures creativity and leads to breakthrough campaigns that not only meet but exceed client expectations.

Picture a tech firm and a retail company teaming up to develop a new app. Initially, the tech firm's prototype ideas were met with skepticism from the retailer. Instead of retreating in silence, the tech firm organized a workshop to collaboratively address the retailer's concerns. By facilitating this open discussion, they built trust and produced a superior app that reflected both parties' visions. This proactive approach illustrates how trust can transform a partnership and enhance innovation.

Lessons Learned:

1. **Transparency is Vital:** I have found that consistent communication builds trust, fostering a strong relationship that can withstand challenges.
2. **Shared Goals Foster Collaboration:** Discussing common goals from the outset ensures all partners are aligned and working toward a shared vision.

3. **Encouraging Open Dialogue:** Creating a safe space for open discussion nurtures creativity and leads to innovative solutions.

Reflective Challenge: How can your organization improve trust with its partners? Are there communication practices you could adopt to foster stronger relationships?

Building trust requires consistent communication and transparency. You should openly discuss goals, concerns, and expectations from the outset. This openness lays the groundwork for a strong relationship that can withstand the challenges of collaboration.

Balancing Efficiency and Innovation

In *Doing Both: How Cisco Captured a Hybrid World*, Inder Sidhu discusses the delicate balance between efficiency and innovation. While it's essential for organizations to operate smoothly and effectively, they must also be willing to embrace new ideas and approaches. This dual focus allows businesses to adapt to changing markets while still driving growth.

In a rapidly evolving market, organizations must find the right balance between optimizing existing processes and exploring new opportunities. Strategic alliances can provide the support needed to pursue innovation while maintaining operational efficiency.

Consider a tech giant that has established a successful cloud service. While they excel at delivering reliable service, they recognize that customer demands are evolving. By forming a strategic alliance with an innovative startup specializing in artificial intelligence, they can enhance their offerings and remain competitive. This partnership not only introduces new technology to their platform but also fosters a culture of continuous improvement that benefits customers and drives growth.

Lessons Learned:

1. **Agility is Key:** Maintaining flexibility in processes allows organizations to pivot quickly when needed. I have witnessed the benefits of agile methodologies firsthand.
2. **Resource Allocation Matters:** Ensuring that both operational efficiency and innovative pursuits receive attention is vital for sustained success.
3. **Collaboration Fuels Innovation:** Partnering with others can help organizations explore new ideas while maintaining quality and efficiency.

Reflective Challenge: How does your organization balance efficiency with the need for innovation? Are there areas where collaboration could help you explore new opportunities without sacrificing operational excellence?

Embracing this dual focus can lead to more robust and successful partnerships.

Conclusion: The Path Forward

As we wrap up this chapter on the importance of collaboration and innovation in strategic alliances, it's clear that these two elements are not just beneficial, they are essential for driving innovative product development. In today's fast-paced business landscape, where customer needs are constantly evolving, organizations must come together to harness their collective strengths and create products that truly resonate with users.

Collaboration fosters a rich environment for creativity and experimentation. When teams from two different organizations unite or two teams from different groups, they bring diverse perspectives, expertise, and ideas to the table. This diversity is akin to a melting pot of innovation, where unique ingredients blend to create something extraordinary.

Think about it: when a tech startup collaborates with a seasoned manufacturing firm, the result can be groundbreaking products that leverage cutting-edge technology while ensuring top-notch production quality. Together, they can create solutions that not only meet market demands but also exceed customer expectations.

Throughout this chapter, we've seen how open communication, and a collaborative mindset are crucial for successful partnerships. By focusing and establishing a culture where all voices are heard and valued, your organization can unlock the full potential of their collaboration.

I have witnessed that when team members feel empowered to share their ideas, they are more likely to contribute to innovative solutions that drive product development forward. This kind of synergy allows partners to navigate challenges and adapt to changing market conditions, ensuring that they remain competitive.

Moreover, the integration of customer feedback into the collaboration process is vital. By actively seeking insights from end-users, partners can identify pain points and opportunities for improvement, ultimately leading to products that are finely tuned to customer needs. This customer-centric approach not only enhances the product itself but also strengthens the partnership, as both parties work toward a common goal of delivering exceptional value.

As I've illustrated through various case studies, successful collaborations can lead to remarkable innovations. Whether it's the partnership between Spotify and Uber that transformed the ride-sharing experience or a retail chain working closely with a logistics company to enhance delivery times, these alliances demonstrate how powerful teamwork can result in products that truly stand out in the marketplace.

In conclusion, **collaboration is the lifeblood of innovative product development.** By collaborating, creating repeatable assets, sharing knowledge, and actively engaging with customers, your organization can create exceptional products that not only satisfy current demands but also anticipate future needs.

As you reflect on your own strategic alliances, consider how fostering collaboration can unlock new avenues for innovation and lead to groundbreaking developments that enhance your offerings.

In your own partnerships, consider how you can cultivate a culture of collaboration and innovation. Are you actively seeking customer feedback? Are you willing to experiment and adapt?

The journey to success in strategic alliances is paved with open communication, mutual respect, and a shared commitment to delivering exceptional value to customers. As you embrace these principles, remember that the most fruitful collaborations often lead to innovative solutions that benefit both the partners involved and create lasting value for customers.

So, collaborate boldly, and innovate fearlessly—because together, you can achieve remarkable things!

Chapter 8

Overcoming Challenges in Strategic Partnerships

Strategic alliances can be fantastic pathways to success, but like any good adventure, they come with their own set of challenges. In the world of partnerships, common pitfalls such as misaligned goals, poor communication, and cultural differences can trip you up if you're not careful. But fear not! By being aware of these challenges and taking proactive steps to address them, organizations can strengthen their partnerships and pave the way for mutual success, all while keeping customer obsession at the forefront.

Common Pitfalls and How to Avoid Them

Unclear Objectives

First up are unclear objectives. Imagine you and your friends decide to plan a vacation. If one of you dreams of a tropical beach while another envisions a snowy mountain retreat, you might end up feeling a bit lost—like trying to navigate a GPS with a map from the 1980s! Misaligned goals can lead to confusion and frustration.

When entering a partnership, it's vital to establish clear objectives right from the start. Think of it as crafting a shared travel itinerary. What are both parties aiming to achieve? By laying out these goals clearly, you create a solid foundation that helps keep everyone aligned and moving in the same direction.

Customer Obsession Alignment:

Aligning objectives with customer needs ensures that the partnership remains focused on delivering value to customers. For instance, if both partners aim to enhance customer service through their collaboration, they should specify how they plan to achieve this, whether through improved response times, integrated technology, or joint training programs.

A tech startup partnered with a marketing firm but found their objectives misaligned. The startup wanted to develop a new app, while the marketing team was eager to promote an entirely different product. After realizing their misalignment, they started joint planning sessions focused on customer needs, transforming what could have been a chaotic relationship into a collaborative brainstorming hub that prioritized user feedback.

Self-Reflection:

Have you set clear objectives with your partners? What steps can you take to ensure everyone is on the same page?

Cultural Misalignment

Next, let's dive into cultural misalignment. This happens when two organizations come together with very different cultures — like mixing oil and water at a salad bar! If one company thrives on creativity and flexibility, while the other is all about structure and routine, there could be some serious clashes.

To avoid this, it's essential to assess cultural compatibility upfront. Spend some time getting to know each other's values and work styles. It's like checking your compatibility before committing to a long-term relationship—only with fewer awkward dinners!

Customer Obsession Alignment:

Understanding cultural differences can significantly impact how both partners respond to customer feedback and innovation. For example, a partner with a more formal structure may have slower decision-making processes, which could hinder rapid responses to customer needs. On the other hand, a more relaxed culture might be better at brainstorming creative solutions but could lack the rigor needed for execution.

In one partnership, one company embraced a "work in your pajamas" culture while the other insisted on business attire. Their first joint meeting resembled a mismatched fashion show! After a good laugh, they blended their cultures and now celebrate casual Fridays together—Hawaiian shirts and all—while ensuring they maintain a strong focus on how their cultural differences enhance their customer approach.

Self-Reflection:

How can you foster a culture that embraces the strengths of both partners while keeping customer needs at the forefront?

Poor Communication

Let's talk about poor communication — the sneaky villain of partnerships! In any collaboration, adequate communication is crucial. When teams aren't sharing updates and ideas, misunderstandings can spiral out of control faster than a cat chasing a laser pointer.

Establishing clear channels for dialogue is critical. Regular check-ins and updates help keep everyone on the same page and prevent the dreaded "I thought you were handling that!" moments.

Customer Obsession Alignment:

Open communication is essential not just between partners but also with customers. Regularly sharing updates about project developments, customer feedback, and market trends can help partners stay aligned in meeting customer expectations.

Two companies worked on a joint project without communicating regularly. One thought the other was creating marketing materials, while the other assumed it was up to their partners. The result? They launched with a fantastic product but zero promotional support—like throwing a surprise party without the cake! Now, they have weekly coffee chats (with plenty of cake) to ensure they're always aligned, sharing customer insights that keep their strategies relevant.

Self-Assessment:

What communication strategies can you implement to improve dialogue with your partners and enhance your responsiveness to customer needs?

Imbalance in Contributions

Next on our list is imbalance in contributions. If one partner feels they're doing all the heavy lifting while the other is coasting along, dissatisfaction can creep in like a cat burglar in the night. It's vital to ensure both parties feel valued and recognized for their contributions.

To maintain balance, regularly assess who's doing what and ensure expectations are clear from the start. Think of it as a potluck dinner; if one person brings a five-course meal and another just a bag of chips, there's bound to be some grumbling!

Customer Obsession Alignment:

Regular assessments can help ensure that both partners are contributing to initiatives aimed at enhancing the customer experience. If one partner is driving all the customer engagement efforts, while the other is not actively involved, it may lead to uneven results that could impact customer satisfaction.

A friend once shared how their team felt overwhelmed while their partner seemed to float by. After a candid discussion, they decided to revisit the workload distribution. Now they celebrate each milestone together, ensuring everyone feels appreciated—and they never show up empty-handed! By actively sharing successes, they have strengthened their customer outreach and improved engagement.

Reflective Question:

How do you ensure that contributions are balanced and recognized within your partnership, especially when it comes to meeting customer expectations?

Insufficient Commitment

Now, let's chat about insufficient commitment. If one partner isn't fully engaged, it can undermine the whole alliance's effectiveness. It's like being in a relationship where one person is all in while the other just shows up for the free snacks!

Continuous commitment from both parties is necessary for long-term success. Encourage engagement at all levels—from the leadership team to the interns.

Customer Obsession Alignment:

A commitment to the partnership means a commitment to your customers. Engaged partners are more likely to stay attuned to customer feedback, allowing for quick adjustments to strategies based on what customers are saying.

A startup found that their larger partner was only half-heartedly involved. After an open and honest conversations, they established regular strategy sessions to rekindle that commitment. It was like pouring fresh fuel on the fire—now, both sides felt excited and invested in their collaboration! Their renewed energy has translated into faster responses to customer inquiries and improved service delivery.

Self-Reflection:

What strategies can you implement to foster commitment from all partners involved, ensuring that customer needs are prioritized?

Failure to Adapt

Lastly, we have the failure to adapt. In a world that's always changing, the inability to respond to new conditions can make an alliance ineffective. Trying to stick to an outdated plan is like trying to fit a square peg into a round hole—frustrating and ultimately ineffective.

Flexibility is key to long-term success. Embrace the lessons learned from both victories and setbacks and be willing to adjust strategies as needed.

Customer Obsession Alignment:

An adaptable partnership can quickly pivot in response to changing customer preferences or market conditions. By regularly reviewing goals and strategies, partners can ensure they remain relevant and valuable to their customers.

I've heard of an organization that launched a partnership but stubbornly stuck to their original plan, even as market conditions shifted. Eventually, they found themselves scrambling to catch up. Now, they hold quarterly reviews to encourage innovation and adaptability, ensuring they can pivot together and maintain their focus on delivering exceptional customer value.

Self-Reflection:

Are you and your partners prepared to adapt when circumstances change? How can you foster a culture of flexibility to better serve your customers?

Lessons on strategic alliances: Stories of Failure and Success

To illustrate these points further, let's look at some real-world examples of strategic alliances—both those that stumbled and those that flourished.

Stories of Failure – where strategic alliance did not work!

Disney and Pixar

In the late '90s, Disney and Pixar formed a partnership that led to blockbuster films like *Toy Story*. However, things took a turn when Disney attempted to renegotiate the terms of their distribution deal. Pixar felt undervalued, and tensions rose. The conflict culminated in a split, with Pixar taking its talents elsewhere and finding immense success on its own. The key lesson here is that aligning objectives and valuing contributions is critical to maintaining partnerships.

Starbucks and Barnes & Noble

Starbucks partnered with Barnes & Noble to create coffee shops within their bookstores, which initially seemed like a match made in heaven. However, as both businesses grew, Starbucks became increasingly focused on its standalone coffee shops. The partnership diluted, and many Barnes & Noble locations faced difficulties as a result. The lesson? Cultural alignment matters. When one partner's growth outpaces the other, it can lead to resentment and misalignment.

Sony and Ericsson

Sony and Ericsson formed a joint venture in 2001 to create mobile phones. While the initial idea seemed promising, cultural differences and management conflicts caused significant setbacks. As they struggled to reconcile their distinct approaches, they ultimately decided to dissolve the partnership in 2012. The lesson learned - understanding and respecting cultural differences can help avert disaster.

Stories of Success – where strategic alliance worked well!

Spotify and Uber

Spotify partnered with Uber to enhance the customer experience by allowing riders to control the music during their ride. This collaboration was a win-win, as it provided Uber with a unique offering while boosting Spotify's visibility. The key takeaway? Clear alignment of goals and a focus on customer experience can yield innovative solutions.

Apple and IBM

In a strategic partnership aimed at enterprise mobility, Apple and IBM combined their strengths to create business apps that leverage the power of iOS. This collaboration allowed both companies to reach new customers and provided innovative solutions to business problems. Their commitment to a shared vision and open communication led to a successful alliance. The lesson here is that aligning objectives with customer needs can drive meaningful innovation.

Nike and Apple

The partnership between Nike and Apple led to the creation of the Nike and Apple iPod Sports Kit, allowing users to track their workouts. This collaboration was driven by a shared vision of enhancing the customer experience for fitness enthusiasts. By combining their technologies, they provided a valuable service that resonated with consumers. The key lesson is that understanding and responding to customer preferences can lead to successful collaborations.

Summary of Lessons Learned

Reflecting on these stories, both the failures and successes, we see several recurring themes:

- **Clarity of Objectives:** Whether a partnership succeeds or fails often hinges on clearly defined goals. Those who took the time to align their objectives with customer needs flourished, while those who neglected this crucial step stumbled.

- **Cultural Compatibility:** Organizations that assessed cultural differences upfront tended to create stronger alliances, whereas mismatched cultures led to tension and conflict.
- **Open Communication:** Partnerships that emphasized transparent communication were better equipped to navigate challenges, while poor communication resulted in misunderstandings and misalignment.
- **Balanced Contributions:** Partnerships that ensured all parties felt valued and recognized thrived, while those with imbalanced contributions faced resentment and disengagement.
- **Commitment and Adaptability:** A continuous commitment to the partnership and a willingness to adapt to changing conditions are essential for long-term success.

By learning from these examples and reflecting on the challenges within your own partnerships, you can foster strategic alliances that not only withstand the test of time but also create meaningful value for your customers.

The secrets to making strategic alliances thrive lie in commitment to collaboration and trust building. When both parties are dedicated, trust flourishes, and flexibility becomes a powerful tool to navigate the challenges that often lead to failure.

Remember, partnerships are like a dance. Sometimes you lead, sometimes you follow, but it's all about moving together in harmony. So, let's embrace the journey, tackle these challenges head-on, and watch our partnerships thrive!

"The whole is greater than the sum of its parts." — Aristotle

Chapter 9

Eight Pillars of Success for Strategic Alliances

As I reflect on the journey of building strategic alliances, I am continually reminded of the profound impact that the Eight Pillars of Success can have on not just the partnerships themselves, but on the very essence of an organization's culture and its relationships with customers. These pillars serve as the guiding principles that elevate our collaborative efforts, allowing us to thrive amidst the complexities of the modern business landscape.

As your organization navigates the complexities of a modern marketplace, it's imperative to understand that the foundations of strategic partnerships are built upon certain key ingredients, I have provided you with eight pillars, but these are not the only ones essential for survival and growth. I encourage you to think about them deeply and widely and how they apply in your daily life.

Let's revisit these pillars and explore the valuable lessons learned that can transform your organization and enhance your strategic alliances.

Mutual Benefit

At the foundation of every successful alliance lies the principle of mutual benefit. Imagine a partnership where both parties feel valued, where each organization contributes its strengths to achieve something greater together. This principle is like a beacon of light guiding your collaborative efforts. Reflect on your own partnerships: **Are both you and your partner actively benefiting from the alliance?** If not, it may be time for an honest conversation to recalibrate and ensure that your collaboration continues to thrive.

Shared Vision

Having a shared vision is akin to possessing a well-drawn roadmap for your journey. It provides direction and keeps everyone aligned with what you aim to achieve. Involve all stakeholders in crafting this vision, ensuring everyone feels a sense of ownership and commitment. Ask yourself: **How well do you communicate your shared vision? Are your goals truly aligned?** A clear and inclusive vision fosters unity and motivation, pushing your teams toward success.

Effective Communication

Effective communication is the lifeblood of any strategic alliance. Without it, misunderstandings can derail progress and lead to frustration. Establishing clear channels for dialogue is vital. Reflect on this: **How often do you check in with your partners?** Regular check-ins and open conversations help maintain alignment and trust. Remember, when communication flows freely, everyone is empowered to contribute, leading to stronger collaboration and innovative solutions.

Cultural Compatibility

As Peter Drucker wisely noted, "Culture eats strategy for breakfast." Understanding and embracing each other's organizational cultures can enhance collaboration significantly. Before entering an alliance, take the time to assess cultural compatibility. Ask yourself: **What are the values and work styles of my partners?** Recognizing and honoring these differences can help prevent friction and foster a harmonious working relationship.

Performance Monitoring

Monitoring the performance of your alliance is crucial for staying on track. Setting clear Key Performance Indicators (KPIs) allows both partners to measure success and identify areas for improvement. Consider: **What metrics are you tracking?** Schedule regular reviews to assess whether you're meeting your goals. This commitment to performance fosters a culture of continuous improvement and keeps your partnership thriving.

Commitment to Collaboration

A strong commitment from both parties is vital for any successful alliance. Reflect on this: **How dedicated is your organization to the partnership?** When leaders visibly advocate for collaboration, it inspires teams to invest their efforts fully. Think back to a time when commitment transformed a relationship; this same dedication can propel your strategic alliances to new heights.

Building Trust

Trust is the glue that holds alliances together, fostering a sense of security and encouraging open communication. Building trust requires transparency and accountability. Ask yourself: **How transparent are you with your partners?** By openly discussing goals and expectations, you lay a solid foundation for collaboration. When trust flourishes, partners feel empowered to share ideas freely, leading to innovative solutions and exceptional outcomes.

Embracing Agility

Finally, embracing agility is essential in today's dynamic environment. The ability to pivot quickly in response to changing needs is what keeps alliances relevant and effective. Reflect on your organization: **How adaptable are you?** Flexibility allows you to explore new opportunities while maintaining operational efficiency. Organizations that can swiftly adjust their strategies are well-positioned for success.

Lessons Learned

Through these eight pillars, I have learned invaluable lessons that can benefit both partners and customers alike:

1. **Mutual Benefit:** Always ensure that both parties feel valued; this strengthens the partnership.
2. **Shared Vision:** Involve all stakeholders in crafting a common vision; it fosters commitment and alignment.
3. **Effective Communication:** Prioritize open dialogue and regular check-ins; this enhances trust and collaboration.
4. **Cultural Compatibility:** Take the time to understand and respect each other's cultures; this prevents conflicts and nurtures harmony.
5. **Performance Monitoring:** Regularly assess KPIs; this drives accountability and continuous improvement.
6. **Commitment to Collaboration:** Leadership commitment is crucial; it inspires teams to invest in the partnership.
7. **Trust Building:** Foster transparency and accountability; trust empowers open communication and innovation.
8. **Embracing Agility:** Stay adaptable to changing needs; flexibility keeps alliances relevant and successful.

Concluding Thoughts

As you reflect on your organization, consider how you can strengthen these pillars within your strategic alliances. The journey toward successful partnerships is ongoing and requires your active participation.

Embrace these principles, and you will unlock the full potential of collaboration, creating a brighter future for your organization and its partners. In the next chapter, I will share a vision of strategic partnership that scale and thrives as a dynamic coalition ecosystem.

Chapter 10

Dynamic Coalition of Strategic Ecosystem

Welcome to a pivotal chapter that explores the transformative concept of **Dynamic Coalition of Strategic Ecosystem**! This idea didn't spring up overnight; it is a culmination of insights from influential authors and my personal experiences navigating the complex world of strategic alliances.

Authors like Reid Hoffman, Michael Hinshaw, Clayton Christensen, Eric Ries, and Charles Handy have laid the groundwork for understanding how collaboration can be the key to thriving in today's dynamic business landscape.

My own journey has been shaped by observing successful partnerships and recognizing the profound impact that customer-centric collaboration can have on both businesses and their clients. These experiences have fueled my belief that when organizations come together with a shared vision and commitment to innovation, the possibilities are limitless.

Embracing the Alliance Mindset

At the heart of Dynamic Coalition of Strategic Ecosystem is the mindset described by Hoffman, Casnocha, and Yeh in "The Alliance." They emphasize the importance of aligning individual aspirations with organizational goals. In a coalition ecosystem, organizations must focus not only on their own objectives but also on the collective mission they share with their partners. This requires a commitment to transparency, trust, and continuous communication.

Imagine a technology startup joining forces with an established enterprise. They are not just collaborating on a project; they are embarking on a journey together, where each partner contributes unique strengths and insights. By creating an environment where talent can flourish, organizations can innovate more effectively and respond to customer demands with agility.

Just as a garden thrives with the right conditions, so too do strategic alliances when nurtured with mutual respect and shared vision.

Customer Obsession as a Guiding Principle

Michael Hinshaw and Bruce Harpham, in "Customer Obsession," argue that truly understanding customer needs is not just a tactic; it's a transformative philosophy that should underpin every interaction. Within a Dynamic Coalition Ecosystem, this customer-centric approach is vital. Partners must continuously gather feedback and iterate on their offerings to ensure they are aligned with customer expectations.

Imagine a cloud service provider partnering with a software development firm. By integrating customer feedback into their development cycles, they can create solutions that resonate with users. This focus on customer obsession not only strengthens the partnership but also leads to higher satisfaction and loyalty among customers. Prioritizing the customer experience unlocks new avenues for growth and innovation for all partners involved.

Navigating the Innovator's Dilemma

Clayton Christensen's "The Innovator's Dilemma" discusses the challenges organizations face when new technologies disrupt existing markets. In a Dynamic Coalition Ecosystem, partnerships can collaboratively navigate these challenges by pooling their resources and expertise. This collaborative advantage allows organizations to stay ahead of technological shifts and meet evolving customer needs.

For instance, consider a traditional banking institution partnering with a fintech startup. They can combine their strengths to innovate in the payments space, providing regulatory knowledge and customer trust while leveraging agility and technological expertise. Together, they can develop cutting-edge financial solutions that cater to a tech-savvy consumer base, effectively overcoming the innovator's dilemma.

Continuous Innovation: The Lean Startup Approach

In "The Lean Startup," Eric Ries champions the need for ongoing innovation and rapid iteration. These principles align seamlessly with the framework of Dynamic Coalition of Strategic Ecosystem, where multiple organizations work together to experiment, learn, and evolve their offerings. The lean methodology promotes an environment where partners can test ideas, gather feedback, and pivot quickly based on real-world insights.

Consider a healthcare provider collaborating with a tech firm to create a telemedicine platform. By adopting a lean approach, they can launch a minimum viable product (MVP), gather user feedback, and refine the offering in real time.

This adaptability accelerates their time to market and ensures the final product genuinely meets the needs of users.

Cultivating Trust and Collaborative Advantage

Charles Handy's "Collaborative Advantage" highlights that collaboration can be a more effective strategy than competition. Trust is a critical component of this collaboration. In his work "The Trusted Advisor," David Meister and his co-authors assert that building trust-based relationships is essential for successful partnerships. This principle is central to Dynamic Coalition of Strategic Ecosystem, where open communication and transparency foster collaboration.

Imagine a technology giant partnering with a nonprofit organization dedicated to sustainability. Both parties can leverage their strengths: the tech company's technological prowess combined with the nonprofit's insights into environmental impact. Together, they can create sustainable solutions that benefit both their missions and the broader community.

A New Era of Possibility

The concept of Dynamic Coalition of Strategic Ecosystem represents a significant shift in how organizations approach strategic alliances. By embracing the principles articulated by these thought leaders, we can build a collaborative framework that prioritizes customer needs, encourages innovation, and establishes trust.

Dynamic Coalition of Strategic Ecosystem are not just a novel idea; they represent a revolutionary approach to strategic alliances where organizations coalesce around shared objectives, driven by a commitment to customer obsession and collaborative innovation. Let's explore what this coalition is all about, what it would look like, and how to set it up effectively.

What Is Dynamic Coalition of Strategic Ecosystem?

At its core, a Dynamic Coalition Ecosystem is a collaborative network of organizations that work together to create value for their customers and themselves. It's about forming strategic alliances that transcend traditional boundaries, fostering an environment where innovation flourishes, and customer needs are prioritized.

This coalition brings together diverse entities—from startups to established enterprises, allowing them to leverage each other's strengths, expertise, and resources. The ultimate goal? To co-create solutions that enhance customer experiences and drive mutual growth.

What Are The Functional Areas Of This Strategic Ecosystem?

Dynamic Coalition of Strategic Ecosystem function as collaborative platforms that facilitate the exchange of ideas, resources, and technologies. Here's what they aim to achieve:

1. Innovative Co-Creation

A successful **Dynamic Coalition Ecosystem** operates on the foundational principles of **Innovative Co-Creation**. This involves members coming together to brainstorm, design, and develop new products and services that address specific customer pain points. In today's rapidly changing markets, the ability to innovate collaboratively can mean the difference between thriving and merely surviving.

Imagine a coalition that includes a cloud service provider, a software development firm, and a customer experience consultancy. Together, they analyze data gathered from customer feedback, identifying common frustrations with existing solutions. By pooling their expertise, they embark on a journey to create an integrated platform that not only enhances functionality but also simplifies the user experience.

For instance, let's consider the successful partnership between Apple and app developers. By fostering an ecosystem where developers can create apps tailored to the iOS platform, Apple has continually expanded its offerings, keeping the user experience fresh and engaging. The App Store itself is a testament to the power of innovative co-creation, where countless apps have been developed to meet diverse customer needs, driving Apple's competitive edge in the market.

Through this collaborative process, members of the coalition can experiment with different ideas, share resources, and mitigate risks. This not only leads to the creation of cutting-edge products but also reinforces the notion that no single organization has all the answers. By working together, they can harness the collective creativity and knowledge within the coalition, pushing the boundaries of what's possible.

2. Shared Learning

Shared Learning is another cornerstone of Dynamic Coalition of Strategic Ecosystem. The idea here is that collaboration isn't just about working together on projects; it's also about fostering an environment where knowledge is freely exchanged, and continuous learning is encouraged. In this ecosystem, every member has something to contribute, and every experience—successes and failures alike—becomes an opportunity for growth.

Think of a coalition made up of companies from different sectors, such as technology, healthcare, and finance. Each organization brings its unique insights and expertise to the table. Regular meetings, workshops, and brainstorming sessions allow them to share best practices, discuss industry trends, and learn from one another's experiences.

For instance, during the COVID-19 pandemic, many businesses had to pivot rapidly. A notable example is how various technology companies shared insights on remote work tools. By collaborating, these companies not only improved their products but also supported each other in understanding the challenges faced by their customers.

This culture of shared learning enables organizations to adapt quickly to changes in the market, equipping them with the knowledge to respond effectively to emerging challenges and opportunities.

Additionally, as members learn from one another, they build a strong sense of community and trust within the coalition. This camaraderie enhances collaboration and allows for more open and honest discussions about challenges, ultimately leading to better outcomes for everyone involved.

3. Enhanced Customer Experiences

In a world where customers have countless options at their fingertips, **Enhanced Customer Experiences** has become a non-negotiable requirement for businesses looking to succeed. Dynamic Coalition of Strategic Ecosystem allow organizations to pool their insights and capabilities to create a seamless and personalized journey for customers, one that goes above and beyond their expectations.

Imagine a coalition comprising a cloud service provider, a data analytics firm, and a marketing agency. By integrating their technologies and expertise, they create a holistic solution that not only addresses customer needs but also anticipates them. For example, they might use data analytics to identify customer behaviors and preferences, enabling them to tailor their marketing strategies and product offerings accordingly.

A real-world example can be found in the collaboration between Nike and its technology partners. By creating a connected ecosystem with apps like Nike+, they provide customers with a personalized experience that tracks workouts and offers tailored recommendations. This seamless integration of technology enhances customer engagement and loyalty, proving that when organizations collaborate effectively, they can deliver truly exceptional experiences.

By focusing on the customer journey and leveraging the strengths of each partner, organizations can create a compelling value proposition that resonates deeply with their target audience. This not only drives customer satisfaction but also encourages long-term loyalty, setting the coalition apart from competitors who may not prioritize such an integrated approach.

4. Increased Agility

Increased Agility is essential in today's fast-paced business environment, and Dynamic Coalition of Strategic Ecosystem provides the framework needed to achieve this flexibility. When partners collaborate, they can respond swiftly to market changes, ensuring they remain relevant and competitive.

Consider a coalition formed between a technology firm and a logistics company. When unforeseen disruptions occur, such as supply chain issues or shifts in consumer demand, the agility of their partnership allows them to pivot quickly. They can reassess their strategies, reallocate resources, and develop new solutions to meet changing customer needs.

A pertinent example is the rapid evolution of grocery delivery services during the pandemic. Companies like Instacart partnered with local grocers to expand their services quickly. By leveraging each other's strengths—technology and logistics—they adapted to the surge in demand for home delivery, keeping their businesses afloat while meeting the immediate needs of customers.

In a Dynamic Coalition Ecosystem, agility is not just about reacting to change; it's also about proactively preparing for it. Regularly reviewing performance metrics, staying attuned to customer feedback, and fostering an environment of open communication enable organizations to stay ahead of the curve. This preparedness not only enhances their ability to respond but also positions them as leaders in their respective industries.

Dynamic Coalition of Strategic Ecosystem is a powerful framework for organizations seeking to thrive in a complex and rapidly changing landscape. By embracing the principles of innovative co-creation, shared learning, enhanced customer experiences, and increased agility, companies can unlock extraordinary value and drive meaningful change.

What Are The Recommended Steps To Setup A Dynamic Coalition of Strategic Ecosystem?

Imagine a vibrant ecosystem where organizations thrive together. Picture a cloud service provider collaborating with fintech startups, healthcare companies, and consumer goods manufacturers. Each partner brings unique skills and perspectives to the table, contributing to a collective pool of knowledge and innovation.

Visually, you might see interconnected circles representing each organization, with lines flowing between them to illustrate the exchange of ideas and resources. This interconnectedness fosters a rich tapestry of collaboration, driving initiatives that benefit not just the partners but also the customers they serve.

Steps to Setting Up a Dynamic Coalition of Strategic Ecosystem

Setting up a Dynamic Coalition Ecosystem involves several key steps:

1. **Identify Potential Partners:** Look for organizations that complement your strengths and align with your vision. Consider their market position, culture, and commitment to innovation.
2. **Establish Shared Goals:** Engage in discussions to articulate the shared objectives of the coalition. Ensure that everyone is on the same page regarding what success looks like.
3. **Develop Governance Structures:** Create a framework that outlines roles, responsibilities, and decision-making processes. This structure will help maintain focus and streamline operations.
4. **Foster Open Communication:** Establish channels for regular communication to facilitate collaboration. Encourage transparency and trust among all members.
5. **Encourage Innovation:** Implement processes for brainstorming and co-creation, allowing partners to explore new ideas and develop innovative solutions together.
6. **Measure Success:** Define key performance indicators (KPIs) that align with your coalition's goals. Regularly assess progress and make adjustments as needed.

Benefits of Dynamic Coalition of Strategic Ecosystem

Dynamic Coalition of Strategic Ecosystem offer a wealth of benefits, including:

- **Increased Innovation:** By collaborating, organizations can leverage diverse perspectives and ideas, leading to groundbreaking products and services.

- **Enhanced Customer Focus:** A coalition that prioritizes customer needs can respond more effectively to market demands, improving customer satisfaction and loyalty.
- **Shared Resources:** Partners can pool their resources, reducing costs and increasing efficiency through collaborative efforts.
- **Greater Market Reach:** The coalition can access new markets and customer segments, broadening its impact and influence.
- **Strengthened Relationships:** Collaborating closely fosters trust and loyalty among partners, leading to more durable and successful alliances.

Dynamic Coalition of Strategic Ecosystem represents a significant shift in how organizations approach strategic alliances. By fostering collaboration, prioritizing customer needs, and embracing innovation, we can redefine the future of business relationships.

Let us dive deeper into a critical aspect of governance, you can also use these as an aid to seek out right skills in the people you appoint to lead this function.

The Critical Role of Governance In Dynamic Coalition of Strategic Ecosystem

Effective governance serves as the backbone that holds your strategic ecosystem together. It's not just about processes and structures; it's about fostering a culture of transparency, accountability, and alignment. Imagine governance as the nurturing soil in which the seeds of collaboration are planted, enabling them to grow and thrive. With a strong governance framework, your organization can confidently navigate the complexities of partnerships, ensuring that both parties are aligned in their objectives and can work harmoniously towards shared success.

1. Stakeholder Integration: The Foundation of Partnership Success

The journey toward effective governance begins with stakeholder integration. This step emphasizes the importance of engaging key stakeholders from both organizations early in the process. It's like inviting everyone to the table, ensuring that each voice is heard and valued. By creating joint governance structures that respect the interests of all parties involved, you lay a solid foundation for collaboration.

Reflect on the partnership between pharmaceutical giant GlaxoSmithKline and biotech firm Adaptive Biotechnologies. From day one, they established a joint steering committee, ensuring that both organizations' interests were represented in every major decision. This early investment in stakeholder engagement not only strengthened their governance but also fostered trust and mutual respect. How often does your organization prioritize the integration of stakeholders in your alliances?

Recommended Action:

Actively integrate key stakeholders from both organizations right from the start. Create joint governance structures that honor the interests of all parties involved and develop shared vision statements that resonate deeply across the partnership. Ensure representation from all relevant departments (e.g., leadership, operations, marketing) and document their roles and responsibilities in a governance structure.

2. Cross-Organizational Communication: The Lifeblood of Partnerships

Effective communication is the lifeblood of any partnership, and it is especially vital for governance. It bridges not just departmental silos but organizational divides, creating a seamless flow of information. Think of it as keeping the lines open in a friendship; regular check-ins and honest conversations help prevent misunderstandings and build trust.

Consider implementing an Alliance Communication Matrix (ACM) to ensure information flows effectively across partnership boundaries. This framework can facilitate multi-level channels for dialogue, allowing you to maintain alignment and momentum. Ask yourself: Are your communication channels fostering a culture of openness, or are they contributing to confusion and misalignment?

Recommended Action:

Implement clear communication channels that bridge departmental and organizational divides. Utilize tools like the Alliance Communication Matrix (ACM) to maintain alignment and momentum, ensuring that all partners stay informed and engaged. Finalize the governance document and obtain approval from all key stakeholders, ensuring that everyone is aligned with the established framework.

3. Joint Capability Development: Building Together

A robust governance framework not only supports but actively facilitates joint capability development. This is where organizations come together to create competencies that neither could achieve alone. By systematically assessing each partner's strengths and weaknesses and identifying complementary capabilities, you can foster a collaborative environment that enhances the partnership's effectiveness.

Reflect on a time when your organization leveraged a partner's unique strengths. How did that collaboration impact your overall performance? Are you actively seeking opportunities to develop joint training programs and knowledge-sharing initiatives to bolster your collective capabilities?

Recommended Action:

Commit to joint capability development by systematically assessing each partner's strengths and weaknesses. Identify complementary capabilities and create joint training programs to enhance collaboration and effectiveness. Document the shared vision and objectives, ensuring they are specific, measurable, achievable, relevant, and time-bound (SMART).

4. Realistic Partnership Planning: The Art of the Possible

Governance requires realistic partnership planning that balances ambition with achievability. It involves creating detailed value creation roadmaps with clear milestones and establishing resource commitments that both partners can sustain. Flexibility is key; your planning framework should allow for adaptability in response to changing circumstances.

Imagine you and your partner are navigating a vast ocean. A solid map guides you, but the ability to adjust your sails in response to shifting winds is crucial. Are your partnership plans detailed yet flexible enough to accommodate unforeseen changes?

Recommended Action:

Develop realistic partnership plans that balance ambition with achievability. Establish detailed value creation roadmaps with clear milestones and ensure that both partners commit resources sustainably. Schedule recurring meetings (e.g., weekly or bi-weekly) and define how progress will be communicated, with a goal to maintain consistent updates.

5. Managing Alliance Resistance: Navigating Cultural Waters

Cultural differences can lead to resistance in strategic alliances, making it essential to address these potential clashes head-on. Effective governance involves early identification of cultural friction points, along with the development of integration strategies that respect both cultures.

Consider the importance of creating conflict resolution mechanisms and change acceleration programs. By fostering a culture that values diversity and encourages open dialogue, you can navigate these waters smoothly. How does your organization currently approach cultural compatibility in its partnerships?

Recommended Action:

Identify potential cultural clashes early in the partnership. Develop integration strategies and conflict resolution mechanisms that honor both organizations' values, fostering a collaborative culture.

6. Shared Success Metrics: Measuring What Matters

In the realm of strategic alliances, what gets measured gets managed. Shared success metrics are vital for ensuring that both partners benefit equitably from the collaboration. By developing balanced scorecards that reflect joint objectives, you can create value-tracking mechanisms that drive continuous improvement.

Reflect on the impact of regular performance reviews on your partnership's health. How often do you assess whether you're meeting your shared goals? Are you open to recalibrating your strategies when targets are missed?

Recommended Action:

Create shared success metrics that reflect joint objectives and implement balanced scorecards to ensure equitable benefits for both partners. Conduct regular performance reviews to drive continuous improvement. Establish at least 3-5 KPIs that are specific to the partnership, such as revenue growth, customer satisfaction scores, or market share increase, and set targets for each.

7. Performance Monitoring: Keeping the Alliance on Track

Continuous monitoring is essential for early detection of both problems and opportunities. Regular health checks of the partnership, tracking value creation metrics, and assessing collaboration effectiveness provide valuable insights into the overall health of the alliance.

Imagine your organization as a finely tuned machine. Regular maintenance checks help ensure it runs smoothly. Are you dedicating enough time and resources to monitor your alliance's performance consistently?

Recommended Action:

Conduct regular health checks on the partnership. Track value creation metrics and assess collaboration effectiveness to provide insights that inform timely adjustments. Collect feedback after major milestones and document the lessons learned, ensuring that these insights are incorporated into future initiatives and adjustments to the governance framework.

8. Sustainable Partnership Development: Build for the Long-Term

Long-term success in strategic alliances requires sustainable governance processes. Developing enduring joint processes, creating mechanisms for continuous value creation, and building lasting relationship networks are essential for fostering an environment conducive to ongoing collaboration and mutual growth.

Consider the importance of building a solid foundation for future endeavors. How are you investing in the longevity of your partnerships? Are you actively creating systems that support continued collaboration and growth?

Recommended Action:
Commit to sustainable partnership development by establishing enduring governance processes. Create mechanisms for continuous value creation and foster lasting relationship networks.

9. Joint Learning and Review: Growing Together

A commitment to learning is a cornerstone of effective governance. Conducting regular partnership effectiveness reviews, documenting lessons learned, and sharing best practices across both organizations foster a culture of continuous improvement.

How often does your organization reflect on its partnerships? Are you creating knowledge management systems that capture insights and drive future initiatives?

Recommended Action:
Embrace a culture of learning by conducting regular partnership effectiveness reviews. Document lessons learned and share best practices across both organizations to drive continuous improvement.

10. Strategic Alliance Evolution: The Path Forward

Governance must evolve alongside strategic alliances to remain relevant. Regular strategic reviews and updates, adapting to changing market conditions, and expanding into new value creation opportunities are crucial for sustaining the partnership's success.

As you navigate the complexities of the business landscape, are you prepared to evolve and adapt your alliances? How do you plan to capitalize on new opportunities that arise?

Recommended Action:

Regularly review and adapt your governance structures to remain relevant. Be open to changing market conditions and explore new value creation opportunities to sustain the partnership's success.

Guidance and Concluding Summary:

In summary, effective governance is the linchpin that allows organizations to measure key metrics accurately and optimize their strategic alliances.

Lack of proper governance in strategic partnerships can spell disaster, creating a landscape filled with confusion and frustration. Imagine entering a partnership with high hopes, only to find that your objectives are misaligned, roles are unclear, and communication feels like a game of broken telephone. It’s disheartening when, despite your best efforts, chaos reigns and trust begins to wane. You might feel as though your contributions are overlooked, leaving you disillusioned and questioning the partnership's value.

Without a strong governance framework, it's easy for partners to step on each other's toes or let important tasks fall through the cracks. This is not just a minor inconvenience; it can lead to significant setbacks, causing you to miss opportunities that could have propelled your collaboration forward. Furthermore, without established metrics to gauge success, you might find yourselves lost, unable to pinpoint areas for improvement or celebrate achievements.

The emotional toll of a poorly governed partnership can be profound. You may feel a sense of betrayal when your expectations aren’t met, and your organization may face the painful reality of wasted resources and energy. Ultimately, the absence of effective governance doesn’t just harm the partnership; it undermines the trust and hope that brought you together in the first place.

To establish effective governance for your strategic partnerships and set the foundation for success, let’s embark on a journey together with seven measurable steps that can truly make a difference.

First, I encourage you to **identify key stakeholders**. Take a moment to create a list of essential individuals from both organizations who will play a pivotal role in the partnership. Ensure that you have representation from all relevant departments—leadership, operations, marketing, and more. This diversity is crucial because it allows different perspectives to shape the governance structure. Document their roles and responsibilities so that everyone knows their part in this collaborative effort.

Next, it’s time to **develop a joint governance framework**. Collaboratively design a framework that outlines the decision-making processes, communication protocols, and conflict resolution mechanisms that will guide you. This framework is your partnership's backbone. Make sure to finalize the governance document and obtain approval from all key stakeholders. Everyone must be aligned with this established framework to foster a sense of ownership and commitment.

Now, let's focus on creating **clear objectives and a shared vision**. I suggest facilitating a workshop where you can come together to craft a shared vision statement and set clear, measurable objectives for the partnership. This is your opportunity to dream big! Document this shared vision and objectives, ensuring they are specific, measurable, achievable, relevant, and time-bound (SMART). This clarity will serve as your guiding star.

Once your objectives are set, it's time to **establish communication channels**. Determine how you will communicate throughout the partnership, including regular meetings and reporting structures. I recommend scheduling recurring meetings—perhaps weekly or bi-weekly—to ensure consistent updates. This proactive communication helps maintain alignment and fosters trust.

Next, let's define **key performance indicators (KPIs)** that will measure the partnership's success. Together, identify KPIs that align with your objectives and truly reflect the partnership's goals. Aim for at least three to five KPIs, such as revenue growth, customer satisfaction scores, or market share increase, and set targets for each. These metrics will keep you focused and accountable.

Now, it's crucial to **implement regular performance reviews**. Schedule periodic performance review meetings to evaluate your progress against the established KPIs and objectives. I recommend conducting these reviews at predetermined intervals—perhaps quarterly. Document the outcomes and be open to adjusting your strategies based on the feedback received during these sessions. This reflective practice ensures that you stay on course and make informed decisions.

Lastly, let's foster a **culture of continuous improvement**. Encourage feedback from all team members involved in the partnership and create a platform for sharing lessons learned and best practices. Collect feedback after major milestones and document the insights gained. Ensure these reflections are incorporated into future initiatives and adjustments to the governance framework. This commitment to learning will enhance your partnership's resilience and adaptability.

By implementing these measurable steps, you'll be well on your way to building a robust governance structure that enhances your strategic partnership's effectiveness, fostering collaboration, and creating sustainable value for all stakeholders involved. **What you do today depends on what you will realize tomorrow!**

As we wrap up this chapter on the **Dynamic Coalition of Strategic Ecosystem**, it's clear that the future of strategic alliances holds incredible promise. We've explored how organizations can come together to create a vibrant ecosystem, united by a shared vision of innovation and customer obsession. Imagine the possibilities when businesses collaborate not just to compete, but to co-create solutions that drive real value for their customers. This is the heart of the Dynamic Coalition of Strategic Ecosystem.

So, let's challenge the status quo and reimagine the possibilities of collaboration. The future is bright, and the opportunities are limitless. Together, with your strategic partners, you can unlock the extraordinary potential of the Dynamic Coalition of Ecosystem and pave the way for a more innovative, customer-centric world.

I invite you to use the following questions for self-reflection on how you can embrace such a collaboration within your organization.

Self-Reflection Questions

As you contemplate the concepts presented in this chapter, here are ten reflective questions to help you assess your current partnerships and their alignment with the Collaborative Innovation Ecosystem framework:

1. Are my current partnerships aligned with our strategic goals and customer needs?
2. How can we identify and engage complementary partners to enhance our offerings?
3. What technologies can we leverage to facilitate collaboration and communication?
4. Do we have clear governance structures in place to ensure smooth operations?
5. How frequently do we encourage and implement creative brainstorming sessions within our teams?
6. What feedback mechanisms are we using to gather insights from both partners and customers?
7. Are we responsive and adaptable to changes in market conditions or customer preferences?
8. How can we foster a culture of continuous improvement and innovation within our organization?

9. What successes or failures can we learn from our past collaborations?
10. How can we measure the success of our partnerships to ensure they deliver meaningful value?

By exploring these questions, you can deepen your understanding of how to cultivate a thriving Collaborative Innovation Ecosystem, positioning your organization to succeed in a rapidly changing world. Commit to creating extraordinary value for your customers!

In Chapter 11, I invite you to reflect on the building actionable steps to redefine your approach to strategic alliances and unlock new potential for growth and innovation.

Chapter 11

Self-Assessment Exercise On Strategic Alliances

This self-assessment is designed to help you think strategically and evaluate your current partnerships and envision the possibilities for future collaborations.

Self-Assessment Questions

Let's start by examining how well you understand your customers. Ask yourself:

1. Understanding Your Customers:

- How well do you understand the needs and preferences of your customers?
- What mechanisms do you currently have in place to gather customer feedback?
- Are you integrating customer insights into your strategic planning?

Next, evaluate your current partnerships:

2. Evaluating Current Partnerships:

- Do your current partnerships align with your organization's strategic goals?
- How do you assess the effectiveness of your alliances in delivering value to your customers?
- Are there areas for improvement in your collaboration processes with existing partners?

As you think about innovation, consider:

3. Innovative Mindset:

- How open is your organization to exploring innovative ideas and solutions?
- What steps can you take to foster a culture of innovation among your team and partners?
- Are you actively encouraging brainstorming sessions or idea-sharing among coalition members?

Flexibility is crucial in today's fast-paced environment:

4. **Flexibility and Agility:**
   - How adaptable is your organization to changes in the market or customer demands?
   - Do you have processes in place for quickly pivoting strategies based on feedback or new information?
   - How can you cultivate an agile mindset within your team and partnerships?

Finally, assess how you measure success:

5. **Measuring Success:**
   - What key performance indicators (KPIs) do you currently use to measure the success of your partnerships?
   - How often do you review these metrics and adjust strategies based on the results?
   - Are you capturing qualitative feedback from customers and partners to inform your assessments?

Use The Following Scoring Mechanism

To facilitate your self-assessment, you can rate your responses on a scale of 1 to 5:

- **1**: Not at all
- **2**: Rarely
- **3**: Sometimes
- **4**: Often
- **5**: Always

Add up your scores for all questions to obtain a total score ranging from 15 to 75.

How To Interpret Your Scores

15-30: Significant Improvement Needed

- Your organization may be missing opportunities for customer engagement and collaboration. Consider taking immediate action to address these areas.

31-50: Room for Growth

- While some practices are in place, there's potential for improvement. Identify specific areas to focus on for enhancement.

51-65: Solid Foundation

- Your organization has a good understanding of customer needs and partnerships. Continue refining your strategies to maximize value.

66-75: Exemplary Practices

- Your organization excels in customer understanding and partnership effectiveness. Share your best practices and consider mentoring others in your industry.

Defining the New Art of Possibility

As you reflect on the questions above and score your responses, consider the following steps to redefine your approach to strategic alliances:

Step:1 Cultivate a Customer-Centric Culture:

Shift your organizational mindset to prioritize customer needs in every decision-making process. Involve customers in co-creation and seek their input on product development.

Step:2 Embrace Collaborative Innovation:

Encourage your team to think creatively and explore new ideas that challenge the status quo. Host regular innovation workshops where all partners can brainstorm solutions to shared challenges.

Step:3 Foster Open Communication:

Establish clear channels of communication with partners and customers. Regularly check in, share updates, and create a safe space for honest feedback. Transparency builds trust and strengthens relationships.

Step:4 Invest in Technology:

Leverage technology to enhance collaboration and streamline processes. Implement tools that facilitate real-time communication and data sharing, enabling partners to work together seamlessly.

Step:5 Monitor and Adjust:

Continuously assess the effectiveness of your alliances using KPIs and qualitative feedback. Be willing to adapt your strategies based on what the data reveals, ensuring that you remain aligned with customer needs and market trends.

Visioning Exercise: The New Art of Possible

To inspire deeper thinking about the future of your strategic alliances, engage in this visioning exercise:

Step:1 Imagine the Future:

Picture what a successful Collaborative Innovation Ecosystem looks like for your organization. What does it achieve? How do customers feel about the value being delivered? Visualize the ideal outcomes.

Step:2 Identify Opportunities:

Write down potential opportunities for collaboration that you haven't explored yet. Consider partnerships with organizations outside your immediate industry that could bring fresh perspectives and ideas.

Step:3 Create an Action Plan:

Based on your reflections, develop a plan outlining specific steps you will take to enhance your strategic alliances. Set measurable goals and timelines to keep yourself accountable.

Step:4 Share Your Vision:

Discuss your vision and plans with your team and partners. Engaging others in the conversation fosters collaboration and can lead to new insights and shared commitments.

Conclusion

By engaging in self-assessment and embracing the principles outlined in this guide, you can redefine the possibilities within your strategic alliances. The landscape of collaboration is evolving, and by prioritizing customer obsession and innovative thinking, you can cultivate partnerships that not only thrive but drive significant value for all stakeholders involved.

As you embark on this journey, remember that the art of collaboration is not just about achieving business goals; it's about creating meaningful connections that empower your organization and elevate the experiences of your customers.

In the next chapter, I provide you with key takeaways to enable you to scale new heights in your world of strategic alliances, where customer obsession rules!

Chapter 12

Conclusion and Key Takeaways

First, thank you for reading this book. It is my intent that the points I have elaborated and outlined in this book will serve as a roadmap for your organization to think and act strategically and build a robust strategic alliance to gain a competitive advantage in today's fast-paced business environment. By embracing customer obsession and building a keen focus on strategic alliance, you and your organization will be able to enhance customer relationships for sustainable growth and operational excellence.

Here is a summary of points outlined in the earlier chapters.

Position Yourself for Growth – Achieve Competitive Advantage

At its core, these pillars focus on understanding and prioritizing customer needs. When an organization commits to customer obsession, they differentiate themselves from competitors who may be less attuned to the evolving preferences and desires of their clients.

For instance, by implementing **Open Communication**, organizations create transparent channels that encourage customer feedback. This responsiveness not only builds trust but also helps organizations anticipate customer needs before they become mainstream.

Imagine a tech company that actively seeks user input on its software updates. By addressing customer pain points in real-time, the company can launch solutions that resonate deeply with users, gaining market share from competitors who may not be as agile.

This proactive approach enables organizations to stand out in crowded markets, fostering loyalty and attracting new customers.

Manage and Measure The Impact – Quantify and Qualify

To truly understand the impact of these laws, organizations need to establish clear metrics. For example, customer satisfaction scores (CSAT) and net promoter scores (NPS) can provide valuable insights into how well the organization meets customer needs. By tracking these metrics over time, companies can evaluate how effectively they are implementing the laws of customer obsession.

Let's say a retail alliance implements a **Stay Customer-Centric** approach, integrating feedback loops into their operations. By measuring the resulting increase in customer satisfaction and repeat purchase rates, the partners can gauge the effectiveness of their strategies. If they see positive trends, it reinforces the value of prioritizing customer needs, allowing them to adjust and refine their approaches as necessary.

Driving Long-Term Growth

The long-term benefits of these pillars of strategic alliance are profound. By embedding customer obsession into their cultures, organizations lay the groundwork for sustained growth. When both partners in an alliance embrace **Mutual Benefit**, they create a collaborative environment that fosters innovation. This commitment can lead to the development of new products and services that not only meet existing customers' needs but also anticipate future trends.

For example, a partnership between a healthcare provider and a technology firm could result in innovative telehealth solutions that adapt based on patient feedback. As these services evolve, they can capture a larger share of the market, driving revenue growth for both partners. This forward-thinking approach ensures that the alliance remains relevant and continues to thrive in the long run.

Ensuring Sustainability of Business Operations

Sustainability is another key advantage of adhering to these pillars. Organizations that prioritize **Continuous Improvement** and **Adaptability** are better equipped to navigate market changes and challenges. By regularly assessing your operations against customer feedback and industry trends, you and your strategic partners can identify areas for enhancement and make necessary adjustments.

Take, for instance, a strategic alliance in the automotive industry. If both companies focus on **Embracing Innovation** and utilize customer insights to drive product development, they can consistently produce vehicles that meet consumer expectations for safety and technology. This commitment to sustainability not only benefits customers but also ensures that both companies remain competitive and profitable in a rapidly changing landscape.

Conclusion

In conclusion, the pillars outlined in this book are meant to provide organizations with a powerful framework to achieve competitive advantage, measure impact, drive long-term growth, and ensure the sustainability of their business operations.

By prioritizing efforts to build a culture that prioritizes customer needs, your organization can create meaningful connections that lead to lasting success. As you continue to navigate the complexities of today's market, embracing these recommended principles of customer obsession will set you on a path toward innovation and excellence. **Are you ready to adopt a customer-obsessed mindset and transform your strategic alliances for the better?**

Imagine a world where collaboration reigns supreme, where communication flows seamlessly, and where every stakeholder is united by a shared commitment to mutual success. This is not merely a vision; it is a possibility that lies within your grasp.

When you prioritize collaboration, you create an environment where ideas can flourish, and innovation becomes the norm. Picture your team, not as isolated entities, but as an intricate tapestry of diverse talents, perspectives, and experiences. Each thread woven together strengthens the fabric of your partnership, making it resilient and adaptable to change. In this collaborative spirit, challenges transform into opportunities, and setbacks become stepping stones to greatness.

By fostering open dialogue, you invite creativity and transparency into the conversation. You empower your partners to voice their thoughts and concerns, cultivating a sense of belonging and a mutual goal and ambition. When everyone feels heard and valued, the foundation of your partnership becomes unshakeable, ready to withstand the storms of uncertainty.

But above all, it is your unwavering commitment to mutual success that will propel your partnerships to extraordinary heights. This commitment is a promise—one that goes beyond transactions and bottom lines. It is a dedication to understanding your partners' goals, celebrating their victories, and standing by them in times of need. .

When you embody this spirit, you foster an atmosphere of trust and respect, creating an ecosystem where everyone thrives. Your customers will feel this commitment as well; they will sense the genuine care and collaboration that underpins every interaction, and they will be drawn to the meaningful value you provide.

I encourage you to embrace this vision and prioritize collaboration, communication, and a commitment to create a world where meaningful value is not just an aspiration but a reality for all.

Build partnerships that not only endure amidst change but flourish in the face of adversity and creates a legacy of shared accomplishments and profound impact.

Recommended Readings

Here are some recommended readings that align with the themes of strategic alliances, customer obsession, and innovation.

These books offer valuable insights and practical guidance to help you deepen your understanding and enhance your approach in these areas:

1. **"The Alliance: Managing Talent in the Networked Age"** by Reid Hoffman, Ben Casnocha, and Chris Yeh

2. **"Customer Obsession: How to Deliver Customer-Centric Experiences"** by Michael Hinshaw and Bruce Harpham.

3. **"The Innovator's Dilemma: When New Technologies Cause Great Firms to Fail"** by Clayton M. Christensen.

4. **"The Lean Startup: How Today's Entrepreneurs Use Continuous Innovation to Create Radically Successful Businesses"** by Eric Ries.

5. **"Doing Both: How Cisco Captured a Hybrid World"** by Inder Sidhu.

6. **"Collaborative Advantage: How Collaboration Beats Competition as a Strategy for Success"** by Charles Handy.

7. **"Customer Experience 3.0: High-Profit Strategies in the New Economy"** by John A. Goodman.

8. **"Trust-Based Selling: Using Customer Focused Solutions to Gain Lifetime Customers"** by Charles H. Green and Andrea P. Howe.

9. **"The Art of Strategic Partnering: Dancing with Elephants"** by Mark Soares.

10. **"The Trusted Advisor"** by David H. Meister, Charles H. Green, and Robert M. Galford.

About the Author:

Ashish G Bhatnagar is a visionary leader and strategic thinker dedicated to transforming how organizations approach customer-centric business practices. With extensive experience in cultivating successful strategic alliances across various sectors, including technology, finance, and healthcare, Ashish is passionate about redefining collaboration in today's fast-paced market.

As the author of "BEYOND TOMORROW - Customer Obsession In Strategic Alliances," Ashish introduces the innovative concept of **Dynamic Coalition Strategic Ecosystems**—a transformative framework that emphasizes agility, co-creation, and mutual value in strategic partnerships. This approach is rooted in the belief that understanding and addressing customer needs is at the heart of every successful alliance.

With a track record of driving innovation and enhancing customer experiences, Ashish inspires organizations to foster a culture of collaboration and responsiveness. His insights, drawn from working with industry leaders and a commitment to continuous learning, empower businesses to think bigger and bolder about the potential of strategic partnerships.

In addition to being an author, Ashish is a sought-after speaker and C-suite advisor, sharing his expertise on the intersection of customer experience and strategic alliances at conferences and workshops worldwide. His work has influenced countless businesses, encouraging them to embrace customer obsession and leverage it for sustainable growth. When not writing or speaking, Ashish enjoys mentoring emerging leaders and exploring the latest trends in technology and business innovation.

Connect with him to discover how to unlock extraordinary possibilities through customer-obsessed strategic alliances that drive success and enhance satisfaction in a dynamic world.

- **LinkedIn**: https://www.linkedin.com/in/agb1388/
- **Email:** tslxllc@gmail.com

Made in United States
Cleveland, OH
04 November 2026

47644441R00056